CONFINED FLUID PHASE BEHAVIOR AND CO₂ SEQUESTRATION IN SHALE RESERVOIRS

CONFINED FLUID PHASE BEHAVIOR AND CO$_2$ SEQUESTRATION IN SHALE RESERVOIRS

YUELIANG LIU
China University of Petroleum, Beijing, China

ZHENHUA RUI
China University of Petroleum, Beijing, China

Gulf Professional Publishing
An imprint of Elsevier

Gulf Professional Publishing is an imprint of Elsevier
50 Hampshire Street, 5th Floor, Cambridge, MA 02139, United States
The Boulevard, Langford Lane, Kidlington, Oxford, OX5 1GB, United Kingdom

Notices
Knowledge and best practice in this field are constantly changing. As new research and experience
broaden our understanding, changes in research methods, professional practices, or medical
treatment may become necessary.

Practitioners and researchers must always rely on their own experience and knowledge in evaluating
and using any information, methods, compounds, or experiments described herein. In using such
information or methods they should be mindful of their own safety and the safety of others, including
parties for whom they have a professional responsibility.

To the fullest extent of the law, neither the Publisher nor the authors, contributors, or editors, assume
any liability for any injury and/or damage to persons or property as a matter of products liability,
negligence or otherwise, or from any use or operation of any methods, products, instructions, or ideas
contained in the material herein.

ISBN: 978-0-323-91660-8

For information on all Gulf Professional publications
visit our website at https://www.elsevier.com/books-and-journals

Publisher: Charlotte Cockle
Senior Acquisitions Editor: Katie Hammon
Editorial Project Manager: Michelle Fischer
Production Project Manager: Prasanna Kalyanaraman
Cover Designer: Mark Rogers

Typeset by STRAIVE, India

Working together
to grow libraries in
developing countries

www.elsevier.com • www.bookaid.org

Contents

About the author   *vii*

Preface   *ix*

Acknowledgments   *xi*

1.   Introduction   **1**

   1.1 Confined fluid-phase behavior in nanopores   2
   1.2 Adsorption behavior of pure hydrocarbons on shale   3
   1.3 Phase behavior of gas mixtures considering competitive adsorption
          effect   4
   1.4 Interfacial tension of CO_2/CH_4/brine system under reservoir
          conditions   4
   References   6

2.   Confined fluid-phase behavior in shale   **9**

   2.1 Comparison of Peng-Robinson EOS with capillary pressure model with
          engineering density functional theory in describing the phase behavior
          of confined hydrocarbons   11
   2.2 Phase behavior of N_2/n-C_4H_{10} in a partially confined space from
          shale   35
   References   52

3.   Adsorption behavior of reservoir fluids and CO_2 in shale   **57**

   3.1 Competitive adsorption behavior of hydrocarbons and hydrocarbon/CO_2
          mixtures in porous media from molecular perspective   59
   3.2 Determination of the absolute adsorption/desorption isotherms of CH_4
          and n-C_4H_{10} on shale from a nanoscale perspective   74
   3.3 Absolute adsorption of CH_4 on shale with the simplified local density
          theory   101
   3.4 Determination of the absolute adsorption isotherms of CH_4 on shale with
          low-field nuclear magnetic resonance   122
   Appendix 1: Position-dependent equation of state parameter [$a_{ads}(z)$]   142
   References   143

4. Interfacial tension for CO_2/CH_4/brine systems under reservoir conditions **151**

4.1 ADSA IFT apparatus 156
4.2 Mathematical formulation 159
4.3 Effect of pressure, temperature and salinity on IFT 162
4.4 Effect of CO_2 concentration on IFT 167
4.5 IFT modeling for CO_2/CH_4/H_2O and CO_2/CH_4/brine systems 173
References 181

5. Oil/gas recovery and CO_2 sequestration in shale **187**

5.1 Selective adsorption of CO_2/CH_4 mixture on clay-rich shale using molecular simulations 187
5.2 Comparing the effectiveness of SO_2 with CO_2 for replacing hydrocarbons from nanopores 206
References 219

6. Summary and commendations **227**

6.1 Summary of this book 227
6.2 Suggested future work 227

Index *229*

About the author

Yueliang Liu is an associate professor at China University of Petroleum (Beijing). His research interests are focused on utilization of CO_2 for enhanced oil-in-place (OIP) recovery and CO_2 storage. He has published more than 50 papers. Yueliang Liu is a member of the Society of Petroleum Engineers (SPE). He also serves as a member of the SPE Production and Facilities Advisory Committee.

Zhenhua Rui is Distinguished Chair Professor, Associate Dean of the College of Carbon Neutrality Future Technology, and Director of CCUS program at the China University of Petroleum (Beijing). His research interests include CCUS, Resource Management, reservoir engineering, petroleum project management, and evaluation. He was a Research Scientist at the Massachusetts Institute of Technology (MIT). He is a member of the United Nations Expert Group on Resource Management, SPE advisory committee member, Hoover Medal Board member, etc. He has published over 80 papers in the energy field. He is a recipient of the SPE International Outstanding Service Award, SPE Distinguished Member Award, SPE Technical Award, SPE Peer Apart Award, and Albert Nelson Marquis Lifetime Achievement.

Preface

A theory is the more impressive the greater the simplicity of its premises, the more different kinds of things it relates, and the more extended its area of applicability. Therefore, the deep impression that classical thermodynamics made upon me. It is the only physical theory of universal content which I am convinced will never be overthrown, within the framework of applicability of its basic concepts.

Albert Einstein

Compared to conventional reservoirs, shale generally consists of a large proportion of organic matter. Organic matter is mainly comprised of kerogen, within which a significant amount of nanopores may reside. Due to the presence of kerogen, the distribution of fluid molecules in shale can be strongly affected by the fluid/pore wall interactions, leading to significant fluid adsorption on the pore surface, thus resulting in a quite different phase behavior in shale reservoirs from that in conventional ones. In view of the characteristics of shale oil, shale reservoirs are recognized as a kind of geological body with great potential for CO_2 storage. In addition, CO_2 is also proposed as a potential agent that can be injected into shale reservoirs for shale hydrocarbon recovery due to its nature at super critical state. Understanding of phase behavior, adsorption behavior, and interfacial properties of fluids in shale is of critical importance for more accurately determining the macroscopic and microscopic distribution of fluids in shale reservoirs as well as understanding the mechanisms governing the CO_2 storage in shale reservoirs. This book presents comprehensive work in understanding the confined fluid–phase behavior and CO_2 sequestration in shale reservoirs.

Chapter 1 provides a brief literature review on confined fluid–phase behavior in nanopores, adsorption behavior of pure hydrocarbons on shale, phase behavior of gas mixtures considering competitive adsorption effect, and interfacial tension of $CO_2/CH_4/brine$ systems under reservoir conditions.

Chapter 2 presents the basic concept of confined fluid–phase behavior in shale. Experimental and theoretical methods are proposed to investigate the confined fluid–phase behavior in shale. The phase behavior of fluids in nanopores from Peng-Robinson equation of state (PR-EOS) with capillary effect is compared with that from engineering density functional theory

(DFT). A new experimental method is proposed to measure the bubble-point pressures of fluids in the presence of actual shale samples.

Chapter 3 describes the fundamental mechanism of the adsorption behavior of reservoir fluids and CO_2 in shale and presents the basic concepts of excess adsorption and absolute adsorption among other things. An efficient numerical solution is proposed to convert the measured excess adsorption to absolute adsorption. This chapter also proposes a novel experimental method to measure the absolute adsorption of shale fluids directly.

Chapter 4 employs the axisymmetric drop shape analysis (ADSA) method to measure the interfacial tension (IFT) between CO_2/CH_4 mixtures and brine over a wide temperature and pressure range. Some key factors, such as salinity, pressure, temperature, and composition, influencing IFT are comprehensively discussed in this chapter.

Chapter 5 presents the fundamental mechanism of shale hydrocarbon recovery and CO_2 sequestration in shale reservoirs. This chapter provides the guidelines for future optimization design of CO_2 injection for shale hydrocarbon recovery and CO_2 sequestration in field applications.

Chapter 6 summarizes the work described in this book and emphasizes the significance in understanding the adsorption behavior and phase and interfacial properties of fluids in shale reservoirs for enhanced shale resources recovery and CO_2 sequestration.

Yueliang Liu
China University of Petroleum, Beijing, China

Acknowledgments

We express our sincere appreciations to Dr. Huazhou Andy Li (University of Alberta), Dr. Ryosuke Okuno (University of Texas at Austin), Dr. Zhehui Jin (University of Alberta), Dr. Hucheng Deng (Chengdu University of Technology), Dr. Yuanyuan Tian (Chengdu University of Technology), and Dr. Xiaomin Ma (Taiyuan University of Technology) for their valuable suggestions and excellent comments to this book.

CHAPTER ONE

Introduction

Shale oil and gas resources are becoming increasingly important since oil/gas production from conventional reservoirs is declining rapidly. The phase behavior of shale fluids is a subject of fundamental importance to shale oil/gas extraction, and the knowledge on the phase behavior of shale fluids is required for the design and optimization of shale oil/gas extraction. For example, as for a shale-gas condensate reservoir, it is important to accurately predict how much condensate will be dropping out at given pressure and temperature as the resulting two phase gas/liquid flow will be fundamentally different from the single gas phase flow (Tan and Piri, 2015). However, phase behavior of shale fluids can be very complex under reservoir conditions due to the strong fluid/surface interactions.

It has been found that phase behavior of shale gas deviates from that in bulk due to the high capillary pressure when confined in nanopore spaces (Nojabaei et al., 2013). However, in nanopores, surface adsorption may be significant, and the distribution of molecules in the pore space is heterogeneous. In addition, shale fluids often consist of multiple components. These components can adsorb on shale surface in a selective manner due to their different levels of adsorption capacities (Haghshenas et al., 2014; Wang et al., 2015). Such selective adsorption phenomenon affects how the different components are distributed in the pore spaces, and how they will migrate in the pore spaces. Besides hydrocarbons, shale fluids may also comprise of formation water with certain salinity. Thereby, the interfacial tension (IFT) of gas/water or gas/brine is one of the most important properties affecting the performance of enhanced gas recovery. It significantly affects the movement, phase behavior, and distribution of reservoir fluids in porous media (Danesh, 1998). Specifically, optimum operations of CO_2 flooding and sequestration in oil/gas reservoirs also depend on accurate knowledge of IFT of CO_2/brine systems, which affects the transport properties and capillary-sealing efficiency of CO_2 in the formation (Li et al., 2013; Chalbaud et al., 2006; Aggelopoulos et al., 2010; Shah et al., 2008).

Confined Fluid Phase Behavior and CO₂
Sequestration in Shale Reservoirs
https://doi.org/10.1016/B978-0-323-91660-8.00003-8

1.1 Confined fluid-phase behavior in nanopores

A number of theoretical and computational approaches have been applied to the study of phase behavior of confined fluids. One popular choice is to use Peng-Robinson equation of state (PR-EOS) by combining the capillary effect. The capillary effect relates the pressure difference between two phases (Travalloni et al., 2010a,b; Nojabaei et al., 2013), which can be evaluated from the Kelvin equation by using the Young-Laplace equation. The PR-EOS with capillary pressure model predicts that the bubble point and lower dew point of hydrocarbon mixtures decrease in nanopores, while the upper dew point increases (Nojabaei et al., 2013; Jin and Firoozabadi, 2016b). Although PR-EOS with capillary pressure model has been widely used, it cannot take into account the intermolecular and fluid-surface interactions which play key roles in the phase behavior of nanoconfined hydrocarbons (Jin and Firoozabadi, 2016b). The assumption of phase equilibrium between an ideal gas and an incompressible liquid phase from the Kelvin equation becomes invalid in nanoscale (Tan and Piri, 2015). In addition, the capillary pressure is usually obtained from the bulk interfacial tensions between two phases (Nojabaei et al., 2013). In nanoscale, the interfacial tensions can be very different from the bulk (Bruot and Caupin, 2016; Singh and Kwak, 2007). To improve the capability of PR-EOS with capillary pressure model, Travalloni et al. (2010a,b) employed two parameters describing fluid-surface interactions to study the phase behavior of confined fluids in porous media. Although this model can calculate the capillary condensation in nanopores, its prediction has a large deviation from molecular simulations.

Recently, molecular simulations and theoretical computations based on statistical thermodynamics have been widely used to study the phase behavior of confined fluids. Among them, Grand Canonical Monte Carlo (GCMC) simulations (Neimark and Vishnyakov, 2000; Singh et al., 2009; Wongkoblap et al., 2011; Jin and Nasrabadi, 2016) and engineering density functional theory (DFT) (Li et al., 2014; Jin and Firoozabadi, 2016a,b) are popular choices. These approaches can explicitly consider the intermolecular and fluid-surface interactions from molecular perspective (Lev et al., 1999). Within the framework of GCMC simulations and DFT, the equilibrium properties of confined fluids are determined by the grand potential minimization (Li et al., 2014). These statistical thermodynamic approaches have shown excellent agreement with experimental data on the gas adsorption and interfacial phenomena (Li and Firoozabadi, 2009; Singh et al., 2009; Li et al., 2014; Jin and Firoozabadi, 2016b). Currently, there is no explicit comparison between the statistical thermodynamic based method and the PR-EOS with capillary pressure model on the phase behavior of confined fluids.

1.2 Adsorption behavior of pure hydrocarbons on shale

Adsorption isotherms of pure hydrocarbons are usually measured on shale samples considering that it provides a fundamental database in simulating phase behavior of shale fluids with the adsorption effect. CH_4, known to be the most abundant component in shale gas reservoirs, is mostly studied. Some heavier hydrocarbons, e.g., C_2H_6, C_3H_8, and nC_4H_{10}, can also present in shale fluids with a large quantity, up to 20 vol% (Wang et al., 2015). However, adsorption behavior of these heavier components is scarcely measured. Pedram et al. (1984) measured the adsorption isotherms of C_2H_6, C_3H_8, and nC_4H_{10} on two oil-shale samples and found nC_4H_{10} has the highest adsorption capacity, followed by C_3H_8 and C_2H_6. But it is noted that the oil-shale they used might still have residual oil left in the samples, which can affect the gas adsorption on shale due to solubility effect of various hydrocarbons in shale oil. Therefore, such measured adsorption isotherms could not represent the actual adsorption capacity of gases on shale. Recently, Wang et al. (2015) measured the excess adsorption isotherms of pure CH_4 and C_2H_6 on shale samples. C_2H_6 is shown to have a higher adsorption capacity than CH_4, and Wang et al. (2015) attributed this finding to that C_2H_6 is more apt to get adsorbed on shale samples than CH_4. But this conclusion is made based on the measured excess adsorption isotherms, rather than the absolute adsorption isotherms; excess adsorption isotherms are generally not accurate enough as it neglects the adsorption–phase volume occupied by the adsorbed gas.

As mentioned above, the measured excess adsorption isotherms neglect the adsorption–phase volume and thereby underestimate the total adsorption amount. The density of the adsorption phase is commonly used to correct the excess adsorption isotherms, yielding the absolute adsorption isotherms. In the adsorption phase, gas molecules are in an adsorbed state; to our knowledge, few efforts are dedicated to quantifying the density of the adsorption phase. Previously, constant density values are normally used to pragmatically represent the density of the adsorption phase. Dubinin (1960) suggested that the density of the adsorption phase is a constant value which correlates with the van der Waals constant b. Later, the density of adsorption phase is argued to be equal to the liquid adsorbate density (Menon, 1968; Wang et al., 2016). Li et al. (2002) compared the aforementioned methods and claimed that the density of the adsorption phase is a function of the system temperature, but its value approaches that proposed by Dubinin (1960). Recently, with molecular simulations, Ambrose et al. (2012) suggested that the density of the adsorption phase correlates with the bulk temperature, pressure, and pore size. Actually, fluids in confined space are strongly affected by fluid/pore-surface interactions, especially in shale samples which are usually abundant in nanoscale pores. It is, thereby, of critical importance to precisely capture the density of the adsorption phase in order to more accurately determine the absolute adsorption isotherms.

1.3 Phase behavior of gas mixtures considering competitive adsorption effect

Shale fluids are usually gas mixtures; individual components in shale fluids generally exhibit selective adsorption behavior on shale, while few efforts have been devoted to understanding how the individual components of a gas mixture become selectively adsorbed on shale and how such selective adsorption alters its phase behavior in confined spaces. It is mainly because the measurements of adsorption equilibrium of gas mixtures are difficult to conduct (Walton and Sholl, 2015). Adsorption equilibrium data of gas mixtures are, however, critical to shale hydrocarbon-in-place estimation and the design of adsorption separation (Walton and Sholl, 2015). Therefore, new experimental approaches are requested to be designed to obtain the fundamental data in order to reveal the essential mechanisms of the adsorption effect on phase behavior of gas mixtures.

1.4 Interfacial tension of CO_2/CH_4/brine system under reservoir conditions

Extensive experimental studies have been conducted on pure gas–pure water systems over wide ranges of pressures and temperatures. Most of the existing studies did not address the effects of nonhydrocarbon contaminants on gas/water IFT, especially at high-pressure/temperature reservoir conditions. Moreover, most of the gas/water IFT measurements are only made for the pure hydrocarbon gases, rather than gas mixtures, with water or brine. Ren et al. (2000) measured the interfacial tension of CH_4/CO_2/H_2O

systems. They covered the temperature range of 76.7–211.7°F and pressure range of 145–4351 psia. But the salinity effect on the IFT was not addressed. In shale formations, the presence of salinity can affect the IFT of reservoir fluids to a large extent. It has been recognized that the addition of salts into the aqueous phase can significantly increase the IFT of gas/brine systems (Massoudi and King, 1975; Li et al., 2012a,b). Some of the previous studies attributed the salinity effect to the change of the interface structure: The cations tend to accumulate in the aqueous phase due to the adsorption of the cations on the interface. Another reason causing the IFT increase is the density increase of the aqueous phase because of salt addition. Although extensive studies have been conducted to measure the IFT of CO_2/brine systems (Yang et al., 2005; Bennion and Bachu, 2008; Chalbaud et al., 2010; Bachu and Bennion, 2009), the experimental data for IFT of CH_4/brine mixtures are limited. Meanwhile, experimental data for IFT of CO_2/CH_4/brine mixtures are still scarce at reservoir conditions.

An accurate IFT model is needed to predict the IFT of gas/brine systems under reservoir conditions. Up to now, numerous correlations were proposed, and some of them have been used in commercial reservoir simulators for estimating IFT by petroleum engineering industry. The Parachor model (Weinaug and Katz, 1943; Macleod, 1923) and the scaling law (Lee and Chien, 1984) have gained more use than other predictive methods because of their simplicity (Danesh, 1998). However, both methods are not recommended for the IFT predictions of hydrocarbon/water systems. Massoudi and King Jr. (1974) presented an IFT correlation for pure CO_2/water systems considering pressure and temperature; but it can be only applied at one temperature. Firoozabadi and Ramey Jr. (1988) proposed an IFT model that can predict the IFT of hydrocarbon–gas/water mixtures. Argaud (1992) and Sutton (2009) developed new IFT correlations based on the Firoozabadi and Ramey Jr. (1988) model by considering a broader class of compounds. Argaud (1992) added the ratio of Parachor to molar mass of each compound to the Firoozabadi and Ramey Jr. (1988) correlation as a corrective factor, while Sutton (2009) considered more parameters in the improved correlation. Nonetheless, the predictive capabilities of these improved models are still limited (Johansson and Eriksson, 1974). Bennion and Bachu (2008) presented an IFT correlation between CO_2 and brine as a function of salinity, which predicts the IFT of CO_2/brine systems based on the solubility of CO_2 in brine. However, the correlation of Bennion and Bachu (2008) cannot predict IFT at pressures and temperatures higher than 3916 psia and 257.0°F. Meanwhile, the correlation was developed based on their own measured data, without being validated by other experimental data. Hebach et al. (2002) and Kvamme et al. (2007) presented IFT correlations for CO$_2$/water mixtures considering reservoir temperature, pressure, and density differences of pure component, but excluding the effect of mutual solubility. Furthermore, Li et al. (2012a,b) and Chalbaud et al. (2009) developed correlations for IFT of CO$_2$/brine mixtures. Other methods based on statistical thermodynamics were also applied to predict IFT, such as linear gradient theory (Yan et al., 2001), perturbation theory (Nordholm et al., 1980), density gradient theory (DGT) (Cahn and Hilliard, 1958; Rowlinson, 1979), and integral and density functional theories (Evans, 1979; Almeida and Telo da Gama, 1989; Bongiorno and Davis, 1975). In general, these methods have not been widely used in the petroleum industry likely due to their complexity.

References

Aggelopoulos, C.A., Robin, M., Perfetti, M., Vizika, O., 2010. CO$_2$/CaCl$_2$ solution interfacial tensions under CO$_2$ geological storage conditions: influence of cation valence on interfacial tension. Adv. Water Resour. 33 (6), 691–697.

Almeida, B.S., Telo da Gama, M.M., 1989. Surface tension of simple mixtures: comparison between theory and experiment. J. Phys. Chem. 93, 4132–4138.

Ambrose, R.J., Hartman, R.C., Diaz-Campos, M., et al., 2012. Shale gas-in-place calculations part I: new pore-scale considerations. SPE J. 17 (1), 219–229.

Argaud, M.J., 1992. Predicting the interfacial tension of brine/gas (or condensate) systems. In: Presented at the SCA European Core Analysis Symposium, Paris.

Bachu, S., Bennion, D.B., 2009. Interfacial tension between CO$_2$, freshwater, and brine in the range of pressure from (2 to 27) MPa, temperature from (20 to 125) °C, and water salinity from (0 to 334 000) mg·L^{-1}. J. Chem. Eng. Data 54, 765–775.

Bennion, D.B., Bachu, S., 2008. A correlation of the interfacial tension between supercritical phase CO$_2$ and equilibrium brine as a function of salinity, temperature and pressure. In: Presented at the SPE Annual Technical Conference and Exhibition. Colorado.

Bongiorno, V., Davis, H.T., 1975. Modified van der Waals theory of fluid interfaces. Phys. Rev. A 12, 2213.

Bruot, N., Caupin, F., 2016. Curvature dependence of the liquid-vapor surface tension beyond the Tolman approximation. Phys. Rev. Lett. 116 (5), 056102.

Cahn, J.W., Hilliard, J.E., 1958. Free energy of a nonuniform system. I. Interfacial free energy. J. Chem. Phys. 28, 258.

Chalbaud, C., Robin, M., Egermann, P., 2006. Interfacial tension data and correlations of brine/CO$_2$ systems under reservoir conditions. In: Presented at the SPE Annual Technical Conference and Exhibition, San Antonio.

Chalbaud, C., Robin, M., Lombard, J.M., Martin, F., Egermann, P., Bertin, H., 2009. Interfacial tension measurement and wettability evaluation for geological CO$_2$ storage. Adv. Water Resour. 32, 98–109.

Chalbaud, C., Robin, M., Lombard, J.M., Bertin, H., Egermann, P., 2010. Brine/CO$_2$ interfacial properties and effects on CO$_2$ storage in deep saline aquifers. Oil Gas Sci. Technol. 65, 541–555.

Danesh, A., 1998. PVT and Phase Behaviour of Petroleum Reservoir Fluids (Ph.D. dissertation). Herriot Watt University, Edinburgh.

Dubinin, M.M., 1960. The potential theory of adsorption of gases and vapors for adsorbents with energetically nonuniform surfaces. Chem. Rev. 60 (2), 235–241.

Evans, R., 1979. The nature of the liquid-vapor Interface and other topics in the statistical mechanics of non-uniform, classical fluids. Adv. Phys. 28, 143–200.

Firoozabadi, A., Ramey Jr., H.J., 1988. Surface tension of water-hydrocarbon systems at Reservoir Conditions. J. Can. Pet. Technol. 27, 41.

Haghshenas, B., Soroush, M., Broh, I.I., Clarkson, C.R., 2014. Simulation of liquid-rich shale gas reservoirs with heavy hydrocarbon fraction desorption. In: SPE Unconv. Resour. Conf. Woodlands, Texas, USA.

Hebach, A., Oberhof, A., Dahmen, N., Kogel, A., Ederer, H., Dinjus, E., 2002. Interfacial tension at elevated pressures-measurements and correlations in the water + carbon dioxide system. J. Chem. Eng. Data 47, 1540–1546.

Jin, Z., Firoozabadi, A., 2016a. Phase behavior and flow in shale nanopores from molecular simulations. Fluid Phase Equilib. 430, 156–168.

Jin, Z., Firoozabadi, A., 2016b. Thermodynamic modeling of phase behavior in shale media. SPE J. 21 (1), 190–207.

Jin, B., Nasrabadi, H., 2016. Phase behavior of multi-component hydrocarbon systems in nano-pores using gauge-GCMC molecular simulation. Fluid Phase Equilib. 425, 324–334.

Johansson, K., Eriksson, J.C., 1974. γ and $d\gamma/dT$ measurements on aqueous solutions of 1,1-electrolyte. J. Colloid Interface Sci. 49, 469–480.

Kvamme, B., Kuznetsova, T., Hebach, A., Oberhof, A., Lunde, E., 2007. Measurements and modelling of interfacial tension for water + carbon dioxide systems at elevated pressures. Comput. Mater. Sci. 38, 506–513.

Lee, S.T., Chien, M.C.H., 1984. A new multicomponent surface tension correlation based on scaling theory. In: Presented at the SPE/DOE Improved Oil Recovery Conference. Tulsa.

Lev, D.G., Gubbins, K.E., Radhakrishnan, R., Sliwinska-Bartkowiak, M., 1999. Phase separation in confined systems. Rep. Prog. Phys. 62 (12), 1573.

Li, Z., Firoozabadi, A., 2009. Interfacial tension of nonassociating pure substances and binary mixtures by density functional theory combined with Peng–Robinson equation of state. J. Chem. Phys. 130 (15), 154108.

Li, M., Gu, A., Lu, X., et al., 2002. Determination of the adsorbate density from supercritical gas adsorption equilibrium data. Carbon 41, 579–625.

Li, X., Boek, E., Maitland, G.C., Trusler, J.P.M., 2012a. Interfacial tension of (brines + CO_2): (0.864 NaCl + 0.136 KCl) at temperatures between (298 and 448) K, pressures between (2 and 50) MPa, and total molarities of (1 to 5) Mol·kg^{-1}. J. Chem. Eng. Data 57, 1078–1088.

Li, X., Boek, E., Maitland, G.C., Trusler, J.P.M., 2012b. Interfacial tension of (brines + CO_2): $CaCl_2$(aq), $MgCl_2$(aq), and Na_2SO_4(aq) at temperatures between (343 and 423) K, pressures between (2 and 50) MPa, and molarities of (0.5 to 5) Mol·kg^{-1}. J. Chem. Eng. Data 57, 1369–1375.

Li, Z., Wang, S., Li, S., Liu, W., Li, B., Lv, Q.C., 2013. Accurate determination of the CO_2-brine interfacial tension using graphical alternating conditional expectation. Energy Fuel 28, 624–635.

Li, Z., Jin, Z., Firoozabadi, A., 2014. Phase behavior and adsorption of pure substances and mixtures and characterization in nanopore structures by density functional theory. SPE J. 19 (6), 1096–1109.

Macleod, D.B., 1923. On a relation between surface tension and density. Trans. Faraday Soc. 19, 38–41.

Massoudi, R., King, A.D., 1975. Effect of pressure on the surface tension of aqueous solutions. Adsorption of hydrocarbon gases, carbon dioxide, and nitrous oxide on aqueous solutions of sodium chloride and tetra-n-butylammonium bromide at 25°C. J. Phys. Chem. 79, 1670–1675.

Massoudi, R., King Jr., A.D., 1974. Effect of pressure on the surface tension of water adsorption of low molecular weight gases on water at 25°C. J. Phys. Chem. 78, 2262–2266.

Menon, P.G., 1968. Adsorption at high pressures. J. Phys. Chem. 72, 2695–2696.

Neimark, A.V., Vishnyakov, A., 2000. Gauge cell method for simulation studies of phase transitions in confined systems. Phys. Rev. E 62 (4), 4611–4622.

Nojabaei, B., Johns, R.T., Chu, L., 2013. Effect of capillary pressure on phase behavior in tight rocks and shales. SPE Reserv. Eval. Eng. 16 (3), 281–289.

Nordholm, S., Johnson, M., Freasier, B.C., 1980. Generalized van der Waals theory. III. The prediction of hard sphere structure. Aust. J. Chem. 33, 2139–2150.

Pedram, E.O., Hines, A.L., Cooney, D.O., 1984. Adsorption of light hydrocarbons on spent shale produced in a combustion retort. Chem. Eng. Commun. 27, 181–191.

Ren, Q.Y., Chen, G.J., Yan, W., Guo, T.M., 2000. Interfacial tension of ($CO_2 + CH_4$) +water from 298 K to 373 K and pressures up to 30 MPa. J. Chem. Eng. Data 45, 610.

Rowlinson, J.S., 1979. Translation of J. D. van der Waals' "the thermodynamic theory of capillarity under the hypothesis of a continuous variation of density". J. Stat. Phys. 20, 197–200.

Shah, V., Broseta, D., Mouronval, G., Montel, F., 2008. Water/acid gas interfacial tensions and their impact on acid gas geological storage. Int. J. Greenhouse Gas Control 2, 594–604.

Singh, J.K., Kwak, S.K., 2007. Surface tension and vapor-liquid phase coexistence of confined square-well fluid. J. Chem. Phys. 126 (2), 024702.

Singh, S.K., Sinha, A., Deo, G., Singh, J.K., 2009. Vapor − liquid phase coexistence, critical properties, and surface tension of confined alkanes. J. Phys. Chem. C 113 (17), 7170–7180.

Sutton, R.P., 2009. An improved model for water-hydrocarbon surface tension at reservoir conditions. In: Presented at the SPE Annual Technical Conference and Exhibition. New Orleans.

Tan, S.P., Piri, M., 2015. Equation-of-state modeling of confined-fluid phase equilibria in Nanopores. Fluid Phase Equilib. 393, 48–63.

Travalloni, L., Castier, M., Tavares, F.W., Sandler, S.I., 2010a. Critical behavior of pure confined fluids from an extension of the van der Waals equation of state. J. Supercrit. Fluids 55 (2), 455–461.

Travalloni, L., Castier, M., Tavares, F.W., Sandler, S.I., 2010b. Thermodynamic modeling of confined fluids using an extension of the generalized van der Waals theory. Chem. Eng. Sci. 65 (10), 3088–3099.

Walton, K.S., Sholl, D.S., 2015. Predicting multicomponent adsorption: 50 years of the ideal adsorbed solution theory. AIChE J. 61 (9), 2757–2762.

Wang, Y., Tsotsis, T.T., Jessen, K., 2015. Competitive adsorption of methane/ethane mixtures on shale: measurements and modeling. Ind. Eng. Chem. Res. 54, 12187–12195.

Wang, L., Yin, X., Neeves, K.B., Ozkan, E., 2016. Effect of pore-size distribution on phase transition of hydrocarbon mixtures in nanoporous media. SPE J. 21 (6), 1981–1995.

Weinaug, C.F., Katz, D.L., 1943. Surface tension of methane-propane mixtures. Ind. Eng. Chem. 35, 239.

Wongkoblap, A., Do, D.D., Birkett, G., Nicholson, D., 2011. A critical assessment of capillary condensation and evaporation equations: a computer simulation study. J. Colloid Interface Sci. 356 (2), 672–680.

Yan, W., Zhao, G., Chen, G., Guo, T., 2001. Interfacial tension of (methane + nitrogen) +water and (carbon dioxide + nitrogen)+water systems. J. Chem. Eng. Data 46, 1544–1548.

Yang, D.Y., Tontiwachwuthikul, P., Gu, Y.A., 2005. Interfacial interactions between reservoir brine and CO_2 at high pressures and elevated temperatures. Energy Fuel 19, 216–223.

CHAPTER TWO

Confined fluid-phase behavior in shale

In nanopores, the thermodynamic properties of confined fluids can be very different from that in bulk (Alfi et al., 2016; Jin and Firoozabadi, 2016a,b). The saturation points of confined fluids are shifted (Luo et al., 2016a,b,c), and strong fluid-surface interaction may result in significant surface adsorption and inhomogeneous density distributions (Cabral et al., 2005; Li et al., 2014). Advancing the understanding of phase behavior of confined fluids is not only crucial to the shale/tight gas and oil recovery (Civan et al., 2012), but also of fundamental importance to many industrial applications, such as heterogeneous catalysis (Cervilla et al., 1994), pollution control (Volzone, 2007), and separation processes (Basaldella et al., 2007).

A number of theoretical and computational approaches have been applied to the study of phase behavior of confined fluids. One popular choice is to use Peng-Robinson equation of state (PR-EOS) by combining the capillary effect. The capillary effect relates the pressure difference between two phases (Travalloni et al., 2010a,b; Nojabaei et al., 2013), which can be evaluated from the Kelvin equation by using the Young-Laplace (YL) equation. The PR-EOS with capillary pressure model predicted that the bubble-point and lower dew-point of hydrocarbon mixtures decrease in nanopores, while the upper dew-point increases (Nojabaei et al., 2013; Jin and Firoozabadi, 2016a,b). Within the framework of PR-EOS with capillary pressure model, the fluid distribution in nanopore is considered to be homogeneous, and confinement effect is taken into account by considering the capillary pressure. PR-EOS with capillary pressure model cannot consider the intermolecular and fluid-surface interactions which play key roles in the phase behavior of nanoconfined hydrocarbons (Jin and Firoozabadi, 2016a,b). The assumption of phase equilibrium between an ideal gas and an incompressible liquid phase from the Kelvin equation becomes invalid in nanoscale (Tan and Piri, 2015). In addition, the capillary pressure is usually obtained from the bulk interfacial tensions between two phases and the curvature (Nojabaei et al., 2013). In nanoscale, the interfacial tensions can be very different from the bulk (Singh and Kwak, 2007; Bruot and Caupin,

2016). To improve the capability of PR-EOS with capillary pressure model, Travalloni et al. (2010a,b) employed two parameters describing fluid-surface interactions to study the phase behavior of confined fluids in porous media. Although this model can calculate the capillary condensation in nanopores, its prediction has a large deviation from molecular simulations. Despite of above-mentioned deficiencies, PR-EOS with capillary pressure model is still a popular choice among engineers and scientists because it is simple and can be easily incorporated into reservoir simulations. However, the reliability of this approach has not been calibrated by more sophisticated statistical thermodynamic approaches yet.

Recently, molecular simulations and theoretical computations based on statistical thermodynamics have been widely used to study the phase behavior of confined fluids. Among them, grand canonical Monte Carlo (GCMC) simulations (Walton and Quirke, 1989; Neimark and Vishnyakov, 2000; Singh et al., 2009; Wongkoblap et al., 2011; Jin and Nasrabadi, 2016) and density functional theory (DFT) (Li et al., 2014; Jin and Firoozabadi, 2016a,b) are popular choices. These approaches can explicitly consider the intermolecular and fluid-surface interactions from molecular perspective (Lev et al., 1999). Within the framework of GCMC simulations and DFT, the equilibrium properties of confined fluids are determined by the grand potential minimization (Li et al., 2014). These statistical thermodynamic approaches have shown excellent agreement with experimental data on the gas adsorption and interfacial phenomena (Li and Firoozabadi, 2009; Singh et al., 2009; Li et al., 2014; Jin and Firoozabadi, 2016a,b). Currently, there is no explicit comparison between the statistical thermodynamic-based method and the PR-EOS with capillary pressure model on the phase behavior of confined fluids.

In the first section, the PR-EOS with capillary pressure model is compared with the engineering DFT (Li et al., 2014) on the phase behavior of confined hydrocarbons. The phase behavior of pure nC_8 and C_1-nC_6 mixtures is investigated in nanopores. These hydrocarbons are commonly seen in shale gas and oil reservoirs (Gasparik et al., 2012; Luo et al., 2016a,b,c). In previous works (Li et al., 2014; Jin and Firoozabadi, 2016a, b) an engineering DFT with PR-EOS is used to study the hydrocarbon adsorption-desorption isotherms as well as gas sorption in shale media and achieved good agreement with experimental data. Recently, the engineering DFT is calibrated by comparing to GCMC simulations on the saturation properties of confined pure and hydrocarbons mixtures (Jin, 2017). Comparing to GCMC simulations, engineering DFT can significantly reduce the computational cost. For simplicity, a structureless carbon slit-pore model is used to describe nanopores. Carbon surface is oil-wet and can provide underlying mechanisms on the effect of nanoconfinement on the phase behavior of hydrocarbons in shale nanoporous media. By comparing to engineering DFT, the validity of the PR-EOS with capillary pressure model is assessed in describing the phase behavior of confined fluids.

2.1 Comparison of Peng-Robinson EOS with capillary pressure model with engineering density functional theory in describing the phase behavior of confined hydrocarbons

2.1.1 Molecular model and theory

Engineering DFT extends bulk PR-EOS to inhomogeneous conditions by using weighted-density approximation (WDA) (Rosenfeld, 1989). On the other hand, the YL equation is used to describe the capillary pressure in the PR-EOS with capillarity. Two approaches are compared by studying the dew-point and other thermodynamic properties of hydrocarbons under nanoconfinement.

2.1.1.1 Engineering density functional theory

Within the framework of the engineering DFT, the phase behavior of hydrocarbons in nanopores is studied in an open system setting. An open system can freely exchange matter and energy with the outside reservoir of given properties. In engineering DFT, the system is in equilibrium with an infinite fictitious reservoir in which the chemical potential μ and temperature T are fixed. The equilibrium thermodynamic properties such as the adsorption and phase behaviors can be obtained by the minimization of the grand potential functional which is a function of density distributions (Ebner et al., 1976). The grand potential functional $\Omega[\{\rho_k(\mathbf{r})\}]$ of the system is related to the Helmholtz free energy functional $F[\{\rho_k(\mathbf{r})\}]$ by,

$$\Omega[\{\rho_k(\mathbf{r})\}] = F[\{\rho_k(\mathbf{r})\}] + \sum_k \int \rho_k(\mathbf{r})[\Psi_k(\mathbf{r}) - \mu_k]d\mathbf{r} \qquad (2.1)$$

where $d\mathbf{r}$ presents the differential volume; $\rho_k(\mathbf{r})$ is the number density distribution of the component k at the position $\mathbf{r}$; $\Psi_k(\mathbf{r})$ is the solid-surface external potential of the component k at the position $\mathbf{r}$; and μ_k is the chemical potential of component k in bulk (Li and Firoozabadi, 2009). At equilibrium, the grand potential functional is at minimum. In other words, the first-order derivative of the grand potential functional over density distributions is zero,

$$\frac{\delta\Omega[\{\rho_k(\mathbf{r})\}]}{\delta\rho_k(\mathbf{r})} = 0 \tag{2.2}$$

With an accurate representation of the excess Helmholtz free energy functional $F^{ex}[\{\rho_k(\mathbf{r})\}]$, the Euler-Lagrange equation can be yielded by minimizing the grand potential functional,

$$\rho_k(\mathbf{r}) = \exp\left[\beta\mu_k - \beta\Psi_k(\mathbf{r}) - \delta\beta F^{ex}[\{\rho_k(\mathbf{r})\}]/\delta\rho_k(\mathbf{r})\right] \tag{2.3}$$

where $\beta = 1/k_B T$; k_B is the Boltzmann constant, and T is the absolute temperature.

In engineering DFT model, the excess Helmholtz free energy functional is divided into two parts: one is obtained from the PR-EOS (Peng and Robinson, 1976; Robinson et al., 1985), in which the weighted-density approximation (WDA) (Rosenfeld, 1989) is adopted to account for the physical interactions between fluid molecules; the other part is supplemented by the quadratic density expansion to account for the long-range interactions (Ebner et al., 1976; Ebner and Saam, 1977). The detailed expressions for the excess Helmholtz free energy functional can be found in the literature by Li and Firoozabadi (2009).

The chemical potentials of fluids are obtained from the PR-EOS. To obtain an accurate equilibrium density distribution, the dimensionless volume shift parameter (VSP) (Jhaveri and Youngren, 1988) is applied by fitting the equilibrium liquid density at $T = 0.7 T_c$. The parameters used in the PR-EOS can be found in Table 2.1.

The binary interaction coefficient between CH_4 and nC_6 is fixed as 0.005 (Nojabaei et al., 2013). The structureless carbon is used to simulate the pores. In a carbon-slit pore, the density distributions were assumed to only vary in the z direction perpendicular to the solid surfaces, i.e., $\rho_k(\mathbf{r}) = \rho_k(z)$.

Table 2.1 Critical temperature T_c, critical pressure P_c, acentric factor ω, molar weight M_w, VSP, and attraction energy parameter ε_g for C_1, nC_6, and nC_8 in the engineering DFT and PR-EOS with capillary pressure model.

Species	T_c (K)	P_c (bar)	ω	M_w (g/mol)	VSP	ε_g/k_B (K)
C_1	190.56	45.99	0.011	16.04	−0.1533	1178
nC_6	507.40	30.12	0.296	86.18	−0.01478	2765
nC_8	568.70	24.90	0.398	114.2	0.04775	3192

The fluid–surface interactions φ_{wk} are described by the 10-4-3 Steele potentials (Steele, 1973),

$$\varphi_{wk}(z) = 2\pi\rho_w\varepsilon_{wk}\sigma_{wk}^2\Delta\left[\frac{2}{5}\left(\frac{\sigma_{wk}}{z}\right)^{10} - \left(\frac{\sigma_{wk}}{z}\right)^4 - \frac{\sigma_{wk}^4}{3\Delta(0.61\Delta + z)^3}\right] \quad (2.4)$$

where $\rho_w = 114\ \text{nm}^{-3}$ and $\Delta = 0.335\ \text{nm}$. Unlike interactions σ_{wk} and ε_{wk} are computed using the standard Lorentz-Berthelot combining rules: $\sigma_{wk} = (\sigma_w + \sigma_k)/2$ and $\varepsilon_{wk} = \sqrt{\varepsilon_w\varepsilon_k}$ with $\varepsilon_w = 28$ K and $\sigma_w = 0.3345$ nm, respectively. The external potential Ψ_k in a slit pore is expressed as,

$$\Psi_k(z) = \varphi_{wk}(z) + \varphi_{wk}(W - z) \quad (2.5)$$

where W is the slit-pore size.

In engineering DFT calculations, the external potentials for C_1, nC_6, and nC_8 are modeled as one CH_4-wall interaction, sum of two CH_3-wall and four CH_2-wall interactions, and sum of two CH_3-wall and six CH_2-wall interactions, respectively, as done in previous works (Jin and Firoozabadi, 2016a, b; Jin, 2017). The modified Buckingham exponential-6 intermolecular potential is used to describe the energy and size parameters of methyl group ($-CH_3$), methylene group ($-CH_2-$), and CH_4. This force field combined with united atom model has shown excellent agreement with experimental data on interfacial tension of various hydrocarbons (Singh et al., 2009). By using the modified Buckingham exponential-6 intermolecular potential, engineering DFT has excellent agreement with GCMC simulations on the saturation properties and critical points of confined hydrocarbons (Jin and Firoozabadi, 2016a,b; Jin, 2017). The parameters σ_k and ε_k are 0.3679 nm and 129.63 K, respectively, for $-CH_3$ group, 0.4 nm and 73.5 K, respectively, for $-CH_2$ group, and 0.373 nm and 160.3 K, respectively, for $-CH_4$ group.

The average density $\langle\rho_{ave,k}\rangle$ of the component k in pores is given as,

$$\rho_{ave,k} = \frac{\int_0^W \rho_k(z)dz}{W} \quad (2.6)$$

2.1.1.2 PR-EOS with capillary pressure model

Within the PR-EOS with capillary pressure model, the pressure difference between the vapor and liquid phases is described by the capillary pressure

P_{cap}, which is related to the interfacial tension γ, contact angle θ, and pore curvature $1/r$,

$$P_{cap} = \frac{2\gamma \cos\theta}{r} \tag{2.7}$$

It assumes that liquid phase completely wets the surface, and the contact angle is zero. The interfacial tension of pure component was obtained from National Institute of Standards and Technology Chemistry WebBook (NIST, 2017). For mixtures, the interfacial tension is computed by Weinaug and Katz (1943):

$$\gamma = \left[\sum_{i=1}^{2} PAC_i(x_i\rho_L - y_i\rho_V) \right]^4 \tag{2.8}$$

where PAC_i represents the Parachor number of component i; ρ_L and ρ_V denote molar densities of liquid and vapor phases, respectively; and x_i and y_i are the mole fractions of component i in liquid and vapor phases, respectively. As presented by previous studies (Santiso and Firoozabadi, 2006; Bui and Akkutlu, 2015), the Parachor model is valid only when the pore radius is larger than 5 nm. Thus, in this study, the phase behavior description by the PR-EOS with capillary pressure model is conducted in pores down to 10 nm.

The phase equilibrium is obtained by achieving the equality of fugacity for each component across the vapor/liquid interface. In pores, the phase pressures used for evaluating the fugacities are different due to the presence of capillary pressure. Thereby, the expression at equilibrium is given as:

For a binary mixture,

$$f_i^V(P_V, T, y_i) = f_i^L(P_L, T, x_i), \ i = 1, 2 \tag{2.9}$$

For a pure component,

$$f^V(P_V, T) = f^L(P_L, T) \tag{2.10}$$

where f_i^V and f_i^L are the fugacities of component i in vapor and liquid phases, respectively; T is absolute temperature; and P_V and P_L are vapor and liquid pressures, respectively. Dew-point is identified as the formation of first droplet.

The above equations for the dew-point calculations are solved by the standard negative flash algorithm; the successive substitution (SS) is initially applied to update the *K-values*, and then followed by the Newton iterations for convergence.

2.1.2 Dew-point calculation

For a pure component, phase description is conducted at an isobaric condition. Within the engineering DFT simulations, starting from a sufficiently high temperature, the temperature is gradually lowered to observe the capillary condensation of confined fluids (sudden jump in average densities in pores); at each temperature, fluid configurations in pores at the previous temperature are used as the initial condition. Successive substitution iteration is used to update the density distributions by using the output of Eq. (2.3) as a new input (Li et al., 2014; Jin, 2017). At the first temperature where the initial guess is not available, the bulk density is then used for the initialization.

For the mixtures, phase description is conducted at isothermal conditions. Within engineering DFT, the lower dew-point of confined fluids is calculated by gradually increasing the bulk pressure while using fluid configurations in pores at the previous pressure condition as the initial condition. The successive substitution iteration is used to update the density distributions by using the output of Eq. (2.3) as a new input. At the first pressure, where the initial guess is not available, the bulk density is used for the initialization. The calculation of lower dew-point is started at a sufficiently low pressure and ended at a sufficiently high pressure. Similarly, for upper dew-point calculations, we start from sufficiently high pressure and gradually lower the bulk pressure. The capillary condensation is monitored to detect both lower and upper dew-points. The capillary condensation refers to sudden jump in the averaged density of confined heavier component.

Engineering DFT can reveal the hysteresis in nanopores and the process of increasing/decreasing pressure would provide different results. However, for PR-EOS with capillary pressure model, there is no hysteresis, and the dew-point for the mixture can be obtained either by increasing or decreasing the pressure.

2.1.3 Critical properties of pure components

The critical points of confined pure hydrocarbons are calculated. The critical temperature of the confined pure component T_c^c is estimated by fitting the liquid-vapor coexistence densities to the following scaling law (Rowlinson and Widom, 1982),

$$\rho_L - \rho_V = A\left(1 - \frac{T}{T_c^c}\right)^{\eta} \tag{2.11}$$

where A and η are the fitting parameter and the characteristic exponent, respectively.

The obtained confined critical temperature is then used to calculate the confined critical density (ρ_c^c) with the assumption that the coexisting vapor-liquid densities obey the so-called rectilinear law (Rowlinson and Swinton, 1982),

$$\frac{\rho_L + \rho_V}{2} - \rho_c^c = B\left(1 - \frac{T}{T_c^c}\right) \tag{2.12}$$

where B is the fitting parameter.

Practically, ρ_c^c and T_c^c are regressed by the least-square fit minimizing the average deviation between the above equations and the simulated coexisting vapor-liquid densities. Once these critical properties are obtained, the confined critical pressure P_c^c can then be estimated by extrapolating the Clapeyron plot to $1/T_c^c$ (Singh and Kwak, 2007),

$$\ln P_c^c = C - \frac{D}{T_c^c} \tag{2.13}$$

where C and D are the fitting parameters.

2.1.4 Phase behavior and critical properties of confined pure nC_8

With the engineering DFT, the average density (ρ_{ave}) of pure nC_8 is presented in nanopores of $W = 5$, 8, and 12 nm at $P = 8$ atm in Fig. 2.1. The results are obtained by decreasing the system temperature. For 5-nm nanopore, ρ_{ave} continuously increases as temperature decreases, but for 8- and 12-nm nanopores, ρ_{ave} has a sudden jump from gas-phase-like density to liquid-phase-like density. This behavior is so-called capillary condensation (Li et al., 2014), and the temperature is defined as the dew-point temperature of confined fluids. These two densities are used to represent the vapor-phase and liquid-phase densities in phase coexistence curve. As indicated in Fig. 2.1, the calculated dew-point temperatures of nC_8 in 8- and 12-nm nanopores are around 502.0 K and 495.5 K, respectively, which is about 8.0 and 1.5 K higher than the bulk saturation temperature (494.0 K) as obtained from the PR–EOS (1978). The increase in dew-point temperature is in agreement with the experimental measurements by Luo et al. (2016b).

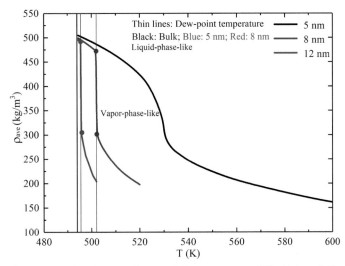

Fig. 2.1 The average density ρ_{ave} of pure nC_8 in nanopores of $W = 5$, 8, and 12 nm at the condition of $P = 8$ atm from the engineering DFT.

For confined pure component, the engineering DFT can calculate the average density in nanopores for the entire temperature range, while the PR-EOS with capillary effect model can have vapor–liquid equilibrium only at the saturation point.

Fig. 2.2 depicts the density-temperature $(\rho - T)$ vapor–liquid coexistence envelopes of nC_8 in nanopores of various pore sizes from the PR-EOS with capillary effect model and the engineering DFT. In engineering DFT, the vapor and liquid phase densities are identified as ρ_{ave} right before and at the capillary condensation as shown in Fig. 2.1. T_c^c obtained from the engineering DFT is shifted to a lower value and approaches T_c as pore size increases, which is in line with previous simulation works (Singh et al., 2009; Didar and Akkutlu, 2013), while no shift is observed from the PR-EOS with capillary effect model. The PR-EOS with capillary effect model predicts that the vapor-phase density in nanopores is similar to that in bulk, while the density obtained from the engineering DFT is significantly higher. While engineering DFT takes into account the fluid–surface interactions and surface adsorptions, the PR-EOS with capillary effect ignores these effects. As a result, the vapor and liquid phase densities obtained from engineering DFT can be higher than that of bulk vapor and liquid phases, respectively. As pore size increases, the PR-EOS with capillary effect predicts that vapor–liquid coexistence curve of confined

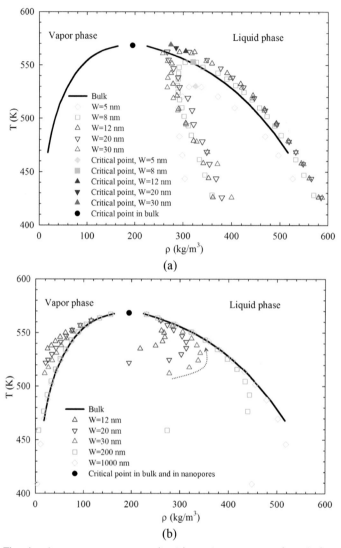

Fig. 2.2 The density-temperature vapor-liquid coexistence curve for nC_8 from (A) the engineering DFT at $W = 5, 8, 12$, and 20 nm and from (B) the PR-EOS with capillary effect at $W = 12, 20, 30, 200$, and 1000 nm. The bulk vapor-liquid density is calculated by the PR-EOS (1978). The filled symbols represent the estimated critical temperatures and densities from Eq. (2.11) to Eq. (2.13). The *dashed arrow* highlights the region where the liquid phase density is lowered due to lower liquid phase pressure from the PR-EOS with capillary effect.

nC_8 approaches bulk. It also predicts that the liquid phase coexistence density first increases as temperature decreases, but after a certain point then decreases. It is because the liquid phase pressure at coexistence is lower than the bulk pressure. As bulk pressure decreases, the capillary pressure increases due to higher interfacial tension. As a result, PR-EOS with capillary effect becomes less reliable at low pressure conditions. Additionally, the PR-EOS with capillary effect model calculates the dew-point with the assumption of vapor-liquid coexistence as a priori. The calculated liquid phase would be in a superheated state with the lower pressure and density. This behavior is completely opposite to the findings from engineering DFT. It implies that stability test with capillary effect should be employed to evaluate phase stability in naopores when using the PR-EOS with capillary effect model, especially in low pressure region (Zhong et al., 2017).

To further investigate the phase behavior of confined fluids, in Fig. 2.3, the dew-point temperatures of pure nC_8 are presented at isobaric conditions of 1, 5, and 20 atm from the engineering DFT, and at 5 and 20 atm from the PR-EOS with capillary effect model. For comparison, the bulk saturation point is depicted by the PR-EOS (1978). From two approaches, as pore size increases, dew-point in nanopores decreases and approaches the bulk

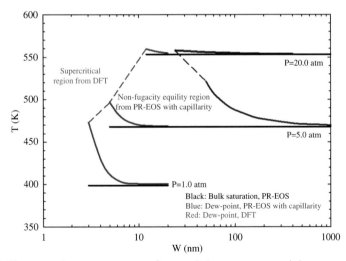

Fig. 2.3 The saturation temperatures of pure nC_8 in nanopores and the corresponding bulk saturation temperatures from the engineering DFT at isobaric conditions of 1, 5, and 20 atm *(red lines)* and the PR-EOS with capillary effect model *(blue lines)* at isobaric conditions of 5 and 20 atm. The *black lines* represent the bulk saturation temperature from the PR-EOS (1978).

saturation temperature. When pore size is smaller than a certain value, there is no capillary condensation from engineering DFT. It indicates confinement–induced supercriticality of hydrocarbons in small nanopores, which is in line with previous experimental measurements (Luo et al., 2016a,b,c). As pressure increases, the supercriticality shifts to larger pores from engineering DFT. The PR-EOS with capillary pressure predicts that when the pore size is sufficiently small, there is no equality of fugacities of vapor and liquid phase. Unlike engineering DFT, the area of nonfugacity equality moves toward a smaller pore size with an increase in pressure. At 5 atm, while engineering DFT predicts that the dew-point temperature of confined nC_8 approaches bulk in 30 nm pores, within the PR-EOS with capillary effect, it is only the case in 1000 nm pores. Zhong et al. (2017) found that the dew-point pressure of confined propane is very close to bulk saturation point when the pore size is around 70 nm. Parsa et al. (2015) claimed that the capillary condensation pressure in 50-nm pores is close to bulk saturation point.

Based on the dew-point temperature calculations, the P-T diagram of nC_8 is depicted in nanopores from the engineering DFT and the PR-EOS with capillary effect model in Fig. 2.4. For comparison, the bulk P-T diagram is presented which is calculated from the PR-EOS (1978). Both approaches predict that as pore size increases, saturation pressure in

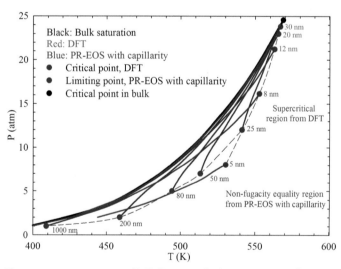

Fig. 2.4 The pressure-temperature (P-T) diagram of nC_8 in nanopores from engineering DFT and PR-EOS with capillary effect model and the corresponding bulk P-T diagram from the PR-EOS (1978).

nanopores increases. At a given system temperature, phase transitions of hydrocarbons in oil-wet nanopores take place at lower pressures than that in the bulk. As for the engineering DFT, fluid-surface interaction may facilitate the phase transitions of confined hydrocarbons. On the other hand, using Monte Carlo simulations, Singh et al. (2009) predicted that in small oil-wet nanopores (pore <2 nm), the saturation pressure of confined hydrocarbons can be higher than the bulk saturation point. Engineering DFT predicts that the P-T diagram of $n\text{C}_8$ approaches bulk for $W = 30$ nm, while the P-T diagram from the PR-EOS with capillary effect model approaches the bulk only when W is as large as 1000 nm. Although the PR-EOS with capillary effect model is in qualitative agreement with the engineering DFT, there is orders of magnitude difference in the critical pore sizes between these two approaches.

While engineering DFT predicts that the confined P-T diagram deviates from the bulk as pressure increases, the opposite is true for the PR-EOS with capillary pressure effect. As pressure increases, the effect of capillary pressure becomes less significant. In addition, the critical pressure from the PR-EOS with capillary effect model does not deviate from the bulk critical pressure. On the other hand, the critical pressure from engineering DFT is lowered in nanopores, which is in line with previous works (Balbuena and Gubbins, 1993; Lev et al., 1999). Due to the lowered critical point, engineering DFT predicts that the supercriticality shifts toward lower pressure and temperature. The supercriticality remains intact for the PR-EOS with capillary effect. We also find that there is nonfugacity equality region from PR-EOS with capillary effect model at low pressure conditions. The PR-EOS with capillary effect model is based on the assumption that confined phase behavior is only affected by capillary pressure, which is dependent on the pore size and surface tension. We use the bulk surface tension for pure $n\text{C}_8$ from the National Institute of Standards and Technology Chemistry WebBook (NIST). However, in nanopores, the surface tension can be very different from that in bulk (Bruot and Caupin, 2016). As suggested by Singh and Kwak (2007), vapor-liquid surface tension under confinement is significantly lower than the bulk value; they proposed that the fluid-pore surface interaction greatly affects the surface tension.

The shifts in the critical pressure are depicted from engineering DFT in terms of $1/W$ in Fig. 2.5. The shift in the critical pressure is calculated from $\Delta P = (P_{cb} - P_{cp})/P_{cb}$, where P_{cb} is the bulk critical pressure and P_{cp} is the confined critical pressure. While the PR-EOS with capillary effect model predicts that there is no deviation on the critical pressure, the engineering DFT

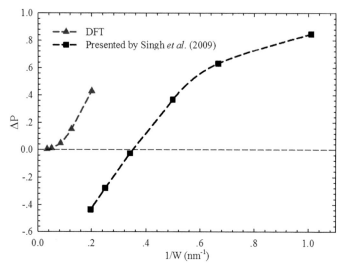

Fig. 2.5 The shift in the critical pressure ΔP of nC_8 vs $1/W$ as calculated by the engineering DFT and the prediction results from Singh et al. (2009).

shows negative deviation. In other words, the critical pressure is lowered in oil-wet nanopores. Singh et al. (2009) studied the shift in the critical pressure of nC_8 with pore confinement in graphite using the configurational-biased grand-canonical transition–matrix Monte Carlo simulations. They observed that when the W is larger than 3 nm, the deviation can be positive, which means that the confined critical pressure increases as illustrated in Fig. 2.5. The shifted critical points by Singh et al. (2009) have been widely used in the so-called EOS modeling with shifted critical properties (Sapmanee, 2011; Devegowda et al., 2012; Alharthy et al., 2013; Jin et al., 2013; Zhang et al., 2013). However, Didar and Akkutlu (2013) have shown that the critical pressure of methane in carbon nanopores decreases by using Monte Carlo simulation. The lower critical pressure is also confirmed by simulation works by Balbuena and Gubbins (1993) and Lev et al. (1999) on Lennard-Jones fluids confined in an attractive pore.

2.1.5 Phase behavior of confined C_1-nC_6 mixture

After studying the phase behavior of pure nC_8 in nanopores, in this subsection, engineering DFT and PR–EOS with capillary effect model are used to investigate the phase behavior of confined C_1-nC_6 mixture. The lower and upper dew-point pressures of this mixture are studied at isothermal conditions. The molar fraction of C_1 in the bulk C_1-nC_6 mixture is fixed at 0.7.

The lower dew-point pressure is first investigated. By increasing the system pressure, the change in the average density (ρ_{ave}) of C_1-nC_6 mixture from the engineering DFT is obtained in nanopores of $W=3$, 5, and 20 nm at three isothermal conditions of $T=410$, 435, and 445 K, respectively, as shown in Figs. 2.6–2.8. Note that $T=445$ K is higher than the bulk cricondentherm $T_{cric}=440.73$ K, which is the highest temperature at which phase transitions can occur in bulk. As bulk pressure increases, the average density of nC_6 shows a sudden jump at the capillary condensation pressure. When $T < T_{cric}$, the lower dew-point pressure with confinement is lower than the bulk and approaches bulk value as pore size increases. In nanopores, the average density of nC_6 has a nonlinear correlation with system pressure. However, the average density of C_1 has a linear behavior. Compared to nC_6, C_1 shows a density drop at the dew-point pressure, as indicated in Figs. 2.6B–2.8B. This opposite behavior is due to the selective adsorption between C_1 and nC_6 molecules; nC_6 molecules have stronger affinity to the pore surface than C_1 molecules. As pore size increases, the average density of nC_6 decreases due to the weaker fluid-surface interactions. On the other hand, for C_1, this is true only when the mixture forms vapor-like structure in nanopores. At $T=435$ K, the average density profile of C_1 and nC_6 in 3-nm nanopore is continuous, and no sudden jump/drop is observed, indicating supercriticality. The same phenomenon is also observed in the 20-nm nanopore at $T=445$ K. In Figs. 2.6–2.8, we also include the results on the variation of the average mixture density with the pore size. As bulk pressure increases, the average mixture density also shows a sudden jump at the capillary condensation pressure; such capillary condensation pressure is the same as that obtained from the average density curves of C_1 and nC_6 of C_1-nC_6 mixture.

In Fig. 2.9, we depict the P-T diagrams of the C_1-nC_6 mixture in nanopores of varying pore sizes in the lower dew-point region from the engineering DFT and the PR-EOS with capillary effect model. For comparison, we also present the bulk P-T diagram from the PR-EOS (1978). When $T < T_{cric}$, the lower dew-point pressures obtained from both engineering DFT and PR-EOS with capillary effect model are shifted to lower pressures. As pore size increases, the confined lower dew-point pressures approach the bulk. When $T > T_{cric}$, phase transition still can occur from the calculations by the PR-EOS with capillary effect model and engineering DFT, which agrees with the previous works (Sandoval et al., 2016). As for the engineering DFT, the highest temperature at which the lower dew-point of confined fluids can be observed is defined as the critical temperature

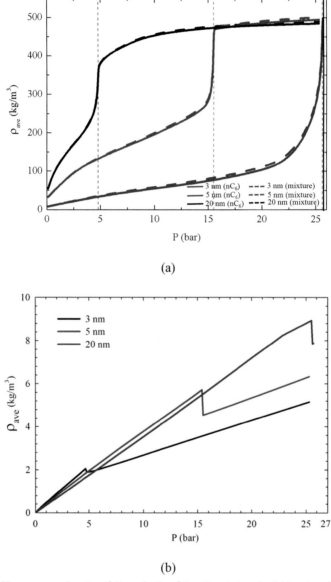

(a)

(b)

Fig. 2.6 The average density of C$_1$ and nC$_6$ of C$_1$-nC$_6$ mixture *(solid lines)* and the average mixture density *(dashed lines)* in nanopores of pore sizes of 3 nm *(black)*, 5 nm *(red)*, and 20 nm *(blue)* at isothermal condition of T=410 K: (A) nC$_6$; and (B) C$_1$. The thin *solid line (black)* presents lower dew-point pressure in bulk; the thin *dash lines (black)*, *(red)*, and *(blue)* present lower dew-point pressures in 3 nm, 5 nm, and 20 nm nanopores, respectively.

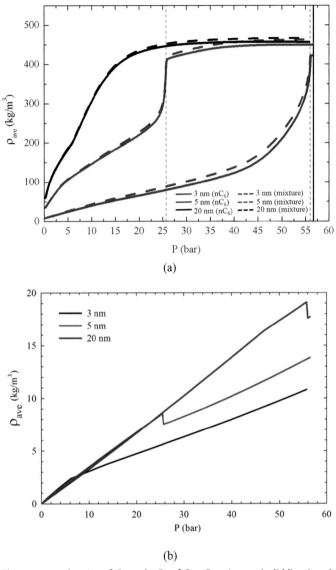

Fig. 2.7 The average density of C_1 and nC_6 of C_1-nC_6 mixture *(solid lines)* and the average mixture density *(dashed lines)* in nanopores of pore sizes of 3 nm *(black)*, 5 nm *(red)*, and 20 nm *(blue)* at isothermal condition of $T=435$ K: (A) nC_6; and (B) C_1. The thin *solid line (black)* presents lower dew-point pressure in bulk; the thin *dash lines (red)*, and *(blue)* present lower dew-point pressures in 5 nm and 20 nm nanopores, respectively.

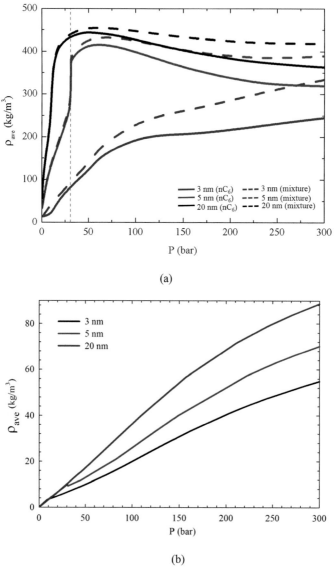

(a)

(b)

Fig. 2.8 The average density of C_1 and nC_6 of C_1-nC_6 mixture *(solid lines)* and the average mixture density *(dashed lines)* in nanopores of pore sizes of 3 nm *(black)*, 5 nm *(red)*, and 20 nm *(blue)* at isothermal condition of $T = 445$ K: (A) nC_6; and (B) C_1. The thin *dash line (red)* presents lower dew-point pressure in 5 nm nanopore.

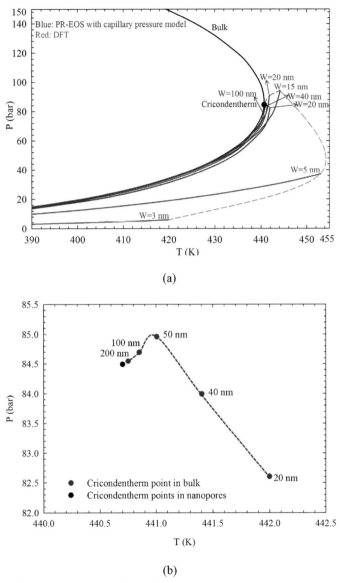

Fig. 2.9 (A) The lower dew-point lines of C_1-nC_6 mixture in nanopores from the PR-EOS with capillary effect model and engineering DFT and the corresponding bulk *P-T* diagram from the PR-EOS (1978); (B) Cricondentherm of the lower dew-point with pore size as obtained from the PR-EOS with capillary pressure model.

of lower dew-point. As shown in Fig. 2.9, the critical temperature of lower dew-point increases till 5 nm and then decreases from the engineering DFT. As for the PR-EOS with capillary effect model, the cricondentherm of lower dew-point decreases with pore size, as presented in the inserted figure in Fig. 2.9. With both approaches, we observe that the critical temperature and the cricondentherm approach the bulk cricondentherm point as pore size increases.

The upper dew-point pressure is then investigated. By decreasing the system pressure, the average density (ρ_{ave}) of C_1 and nC_6 of C_1-nC_6 mixture obtained from the engineering DFT is presented in nanopores of $W=20$, 30, and 50 nm at three isothermal conditions of $T=430$, 435, and 439 K, respectively, as shown in Figs. 2.10A, 2.11A, and 2.12A. Note that these temperatures are all lower than the bulk cricondentherm $T_{cric}=440.73$ K. The desorption isotherms of C_1-nC_6 mixture are obtained by decreasing the system pressure, which represents the calculations of upper dew-point pressures of confined fluids. As bulk pressure decreases, the average density of nC_6 decreases first and then increases. It also shows a sudden jump at the capillary condensation pressure at 439 K for all pore sizes. The upper dew-point pressure with confinement is higher than that in the bulk and approaches the bulk value as pore size increases. As pore size increases, the average density of nC_6 decreases due to weaker fluid-surface interactions. At $T=435$ K, the average density profile of nC_6 in 20-nm pores is continuous, and no sudden jump is observed over the entire pressure range, while in 30-m and 50-nm pores, there is capillary condensation. The noncapillary condensation indicates confinement-induced supercriticality of hydrocarbons in small nanopores. The same phenomenon is also observed in the 20- and 30-nm pores at $T=430$ K. In other words, supercriticality shifts toward higher pore size as temperature drops. Similar to the adsorption process, in nanopores, the average density of nC_6 has a nonlinear correlation with system pressure and the average density of C_1 has a linear behavior during desorption process. Compared to nC_6, C_1 shows a density drop at the lower dew-point pressure, as indicated in Figs. 2.10B, 2.11B, and 2.12B. This opposite behavior can be attributed to the selective adsorption between C_1 and nC_6 molecules; nC_6 molecules have stronger affinity to the pore surface than C_1 molecules.

Fig. 2.13 presents the adsorption and desorption isotherms of nC_6 of C_1-nC_6 mixtures in a nanopores with $W=20$ nm at various temperatures. The confined fluids experience capillary condensation and then evaporation in adsorption and desorption isotherms at $T=441$ and 441.8 K. These temperatures are higher than T_{cric}. Interestingly, the condensation behaviors

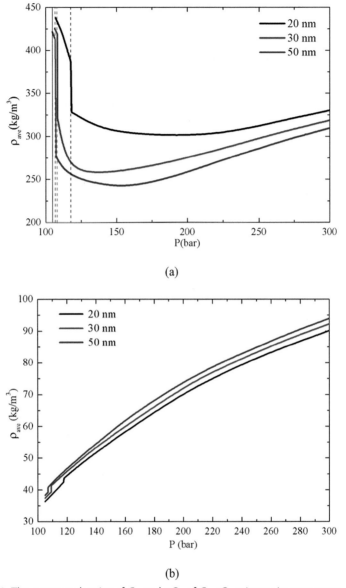

Fig. 2.10 The average density of C_1 and nC_6 of C_1-nC_6 mixture in nanopores of pore sizes of 20 nm *(black)*, 30 nm *(red)*, and 50 nm *(blue)* at isothermal condition of $T = 439$ K: (A) nC_6; and (B) C_1; the thin *solid line (black)* presents upper dew-point pressure in bulk; the thin *dash lines (black)*, *(red)*, and *(blue)* present upper dew-point pressures in 20 nm, 30 nm, and 50 nm nanopores, respectively.

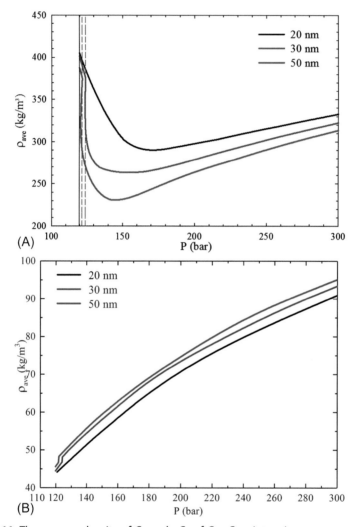

Fig. 2.11 The average density of C_1 and nC_6 of C_1-nC_6 mixture in nanopores of pore sizes of 20 nm *(black)*, 30 nm *(red)*, and 50 nm *(blue)* at isothermal condition of $T = 435$ K: (A) nC_6; and (B) C_1; the thin *solid line (black)* presents upper dew-point pressure in bulk; the thin *dash lines (red)*, and *(blue)* present upper dew-point pressures in 30 nm, and 50 nm nanopores, respectively.

from adsorption and desorption isotherms coincide at around 441.8 K. At $T = 442$ K, the adsorption and desorption isotherms overlap in the entire range, indicating supercriticality. During pressure drop, capillary condensation in nanopores can be understood as retrograde condensation. As a result, such dew-point corresponds to the upper dew-point, which is different from

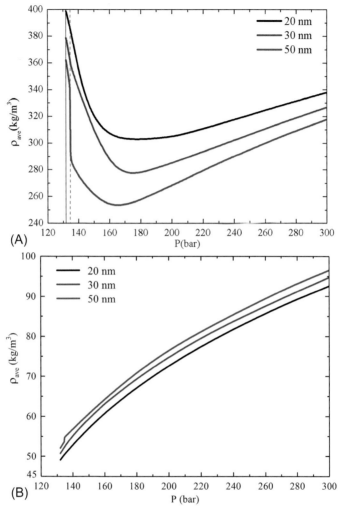

Fig. 2.12 The average density of C_1 and nC_6 of C_1-nC_6 mixture in nanopores of pore sizes of 20 nm *(black)*, 30 nm *(red)*, and 50 nm *(blue)* at isothermal condition of $T=430$ K: (A) nC_6; and (B) C_1; the thin *solid line (black)* presents upper dew-point pressure in bulk; the thin *dash line (blue)* presents upper dew-point pressure in 50 nm nanopores.

the lower dew-point observed from adsorption process. During desorption, after forming liquid-like phases within nanopores, heavier component can have capillary evaporation as pressure further drops. On the other hand, during adsorption, heavier component first experiences capillary condensation and then evaporation as pressure further increases.

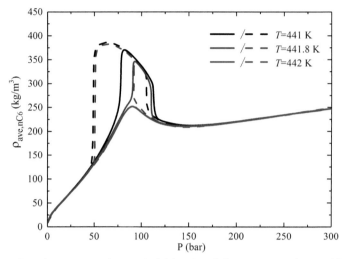

Fig. 2.13 The adsorption isotherms *(solid lines)* and desorption isotherms *(dash lines)* of nC_6 in the C_1-nC_6 mixtures in nanopores with a pore size of 20 nm at various temperatures.

In Fig. 2.14, the *P-T* diagram of C_1-nC_6 mixture is presented in nanopores of varying pore sizes in the upper and lower dew-point regions. Both Engineering DFT and the PR-EOS with capillary pressure model predict that upper dew-point pressure increases. As pore size increases, the upper dew-point of confined fluids approaches the bulk. Interestingly, engineering DFT predicts that the upper dew-point line coincides with the lower dew-point line at one point as shown in Fig. 2.14. This point is defined as the cricondentherm point of confined fluids. As temperature decreases, the confined upper dew-points show departure from the corresponding bulk values. If temperature further drops, there is supercritical region predicted from engineering DFT, where no capillary condensation is observed as pressure drops. On the other hand, the PR-EOS with capillary effect model does not show such supercriticality and the upper dew-point line passes through the bulk critical point due to zero capillary pressure at the critical point.

Fig. 2.15 depicts the deviations of the confined dew-point pressures for the C_1-nC_6 mixture as predicted from the PR-EOS with capillary effect model and engineering DFT. Both approaches predict that deviation of the confined dew-point pressures decreases as pore size increases. For the PR-EOS with capillary effect model, the dew-point pressure difference

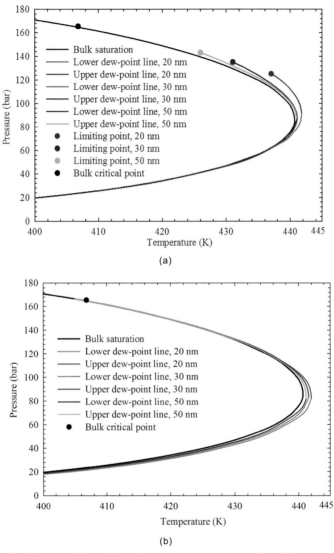

Fig. 2.14 The dew-point lines of C_1-nC_6 mixture in nanopores from (A) engineering DFT and (B) PR-EOS with capillary effect model.

increases as system temperature increases. At the bulk critical temperature, the upper dew-point pressure remains unchanged. However, as illustrated in Fig. 2.15A, engineering DFT predicts that, as temperature increases, the upper dew-point pressure difference decreases first, and then increases in 30- and 50-nm nanopores. Besides, in the vicinity of the upper dew-point,

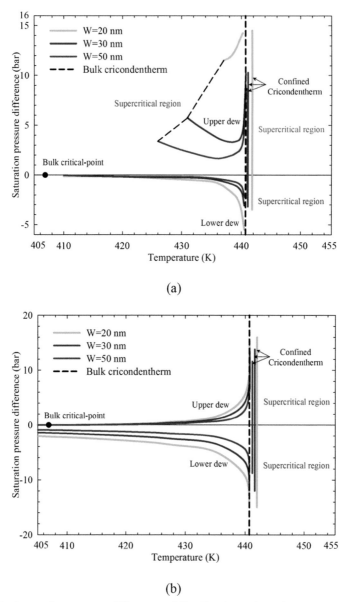

(a)

(b)

Fig. 2.15 Dew-point pressure differences of C_1-nC_6 mixture in various nanopores from (A) engineering DFT and (B) PR-EOS with capillary effect model. The confined cricondentherm correspond to the highest temperatures when phase transition in nanopore occurs.

engineering DFT calculation indicates that, when temperature is lower than a certain value, there is no capillary condensation. As pore size increases, the supercriticality shifts to a lower temperature. Nojabaei et al. (2013) proposed that phase transition cannot happen beyond the cricondentherm. However, both PR-EOS with capillary effect model and engineering DFT predict that phase transitions can still occur at temperatures beyond the bulk cricondentherm point.

Engineering DFT can provide important insights into the phase behavior modeling in unconventional shale reservoir. It can also provide necessary corrections and guidance to the conventional EOS modeling (i.e., correct shifted saturation properties and hysteresis). Engineering DFT can also possibly be coupled into reservoir simulations by obtaining the phase behavior of confined fluids of various composition, pressure, and temperature as priori. Pore size distribution can also affect the phase behavior modeling. There have been some conventional EOS modeling (Wang et al., 2016; Luo et al., 2017) and Monte Carlo (MC) simulations works (Jin et al., 2017) on the effect of pore size distributions on the phase behavior of pure and hydrocarbon mixtures in nanoporous media. However, inhomogeneous density distributions and surface adsorption are still ignored in conventional EOS modeling and MC simulation is greatly hampered by expensive computational cost, especially for heavy hydrocarbons. In addition, based on the SEM imaging, it has been revealed that pore structures in shale may not only be slit-shaped, but also include cylindrical and ink-bottle shapes (de Boer and Lippens, 1964; Sing et al., 2008). The adsorption-desorption isotherms of hydrocarbons in these shaped pores can be different from the slit-shaped. For example, within ink-bottle model, depending on the pore diameter ratio of "ink" and "bottle," nitrogen adsorption-desorption isotherms may behave differently due to pore-blocking and cavitation effects (Fan et al., 2011; Klomkliang et al., 2013). In future, we will also explore the effect of pore size distribution and pore geometry on the phase behavior of nanoconfined pure and hydrocarbon mixtures.

2.2 Phase behavior of N_2/n-C_4H_{10} in a partially confined space from shale

Unlike conventional reservoirs, shale reservoir possesses unique characteristics, such as ultra-low permeability and strong heterogeneity. Nanoscale pores are dominant pores in shale reservoirs, and their diameters are normally between 1 and 20 nm (Nagarajan et al., 2013). Thermodynamics and phase equilibria of fluids in nanopores (called "confined spaces") are more complicated than those in bulk spaces. In such pores, the in situ phase behavior in confined spaces is affected by pore wall-fluid interaction, capillary pressure, and sorption of hydrocarbon on shale material (Li et al., 2014; Jin and Firoozabadi, 2016a,b).

Extensive studies are devoted to understanding the phase behavior of shale fluids. Some modeling studies were conducted to describe phase behavior of shale fluids in confined spaces. Nojabaei et al. (2013) coupled capillary pressure with phase equilibrium relation to describe phase behavior of confined fluids by use of the Peng-Robinson equation (PR-EOS) (Peng and Robinson, 1976). They reported that small pores decreased bubble-point pressures, and either decreased or increased dew-point pressures. Travalloni et al. (2014) modeled phase behavior of confined fluids in homogeneous and heterogeneous porous media using an extended PR-EOS. Their modeling study showed that small pores may confine phases with very similar or very different densities and compositions. Dong et al. (2016) numerically studied phase equilibrium of pure components and their mixtures in cylindrical nanopores using the PR-EOS coupled with capillary pressure and adsorption theory. The reduction of pore diameter was considered in their model due to the existence of adsorption film. Wang et al. (2016) numerically investigated the effect of pore size distribution on phase transition of hydrocarbon mixtures in nanoporous media. They presented a procedure to simulate the sequence of phase transition in nanoporous media, and found that a phase change always occurs firstly in the larger pores, and then in the smaller pores.

Experimental studies on the phase behavior of fluids contained in the nanopore spaces of shale are, however, relatively scarce in the literature. Morishige et al. (1997) measured adsorption isotherms of pure gases on mesoporous MCM-41 molecular sieves with different pore sizes. Their experimental adsorption data showed that the critical temperatures of pure fluids in mesopores were quite different from those in the bulk space. Yan et al. (2013) applied differential scanning calorimetry and temperature-dependent X-ray diffraction technology to obtain phase behavior of n-tridecane/n-tetradecane mixtures in the bulk and in the confined porous glass. The mixtures showed a similar phase behavior to the bulk, especially in larger pores than 30 nm. Under confinement, their phase behavior varied with pore size as well as temperature and composition. Wang et al. (2014) applied nanofluidic devices to visualize phase changes of pure alkane and alkane mixtures under nanoconfinement. The vaporization of liquid phase in nanochannels (5 μm wide by 100 nm deep) was remarkably suppressed in comparison to that in micro-channels (10 μm wide by 10 μm deep). Pure alkanes and alkane mixtures exhibited different vaporization behavior; this was because the liberation of lighter components from the liquid phase to the gas phase in the micro-channels increased the apparent molecular weight of the liquid in the nanochannels, suppressing its bubble point (Wang et al., 2014). Alfi et al. (2016) investigated phase behavior of pure Hexane, Heptane, and Octane inside nanochannels of 50 nm using lab-on-a-chip technology integrated with high-resolution imaging techniques. They found that, in a nanochannel with a width of 50 nm, the confinement effect in the form of wall-molecule interactions was almost negligible. Luo et al. (2016a,b,c) experimentally explored the relationship between saturation temperature and pore diameter for n-hexane, n-octane, and n-decane that were confined in silicate nanoporous materials CPG-35 using differential scanning calorimetry. They observed that the saturation temperature in nanopores was higher than that in a bulk space.

To summarize, extensive studies have been conducted to elucidate the effect of capillary pressure on the phase behavior of confined fluids. However, the effect of sorption on the phase equilibrium of shale fluids is scarcely investigated, although it is an important and common phenomenon in shale reservoirs. To our knowledge, no publications have reported experimental data on the phase behavior of fluids in the presence of real shale materials.

n-C_4H_{10}, as a nonvolatile hydrocarbon, is a common component present in the shale gas-condensate reservoirs; N_2, as a volatile nonhydrocarbon component, can be introduced into the reservoir as an energized fluid used during an energized fracturing treatment. Furthermore, N_2 has been found to be a main component that can be produced from some shale reservoirs, e.g., Antrim shale and Barnett shale. Hence, the N_2/n-C_4H_{10} binary is selected to elaborate the effect of sorption on the in situ phase behavior of shale fluid. This section presents an experimental study on the phase behavior of two binary mixtures in a partially confined space. Fig. 2.16 illustrates the physical model representing a partially confined space, which consists of a bulk space and a confined space. The major objective is to explore the impact of sorption on fluid-phase behavior. Bubble-point pressures for the binary mixtures were measured firstly in a bulk space, and then in a partially confined space. It delineates the reasons leading to differences in bubble-point pressure between the bulk and the partially confined space.

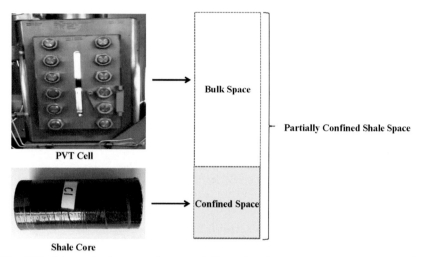

Fig. 2.16 Schematic diagram of the partially confined space that consists of a bulk space and a confined shale space.

To our knowledge, this is the first time that the effect of sorption on the phase behavior of fluid mixtures in the presence of real shale is measured.

2.2.1 Measurement of the P/V isotherms in confined shale

This section describes the N_2 adsorption/desorption test used for characterizing both shale cores (#1 and #2), the procedure of total organic carbon (TOC) measurement for shale cores, the measurement of system volumes, and the constant composition expansion (CCE) method for measuring P/V isotherms. Fig. 2.17 shows the schematic diagram of the experimental setup for measuring P/V isotherms of N_2/n-C_4H_{10} mixtures.

Before the measurement, both shale cores (#1 and #2) were placed in an oven, and vacuumed for 48 h and heated at a constant temperature of 423.15 K with helium as a carrier gas to remove moisture and other adsorbed gases. This pretreatment was called "degassing and drying treatment." Helium within the shale samples was then evacuated before the subsequent steps. After this treatment, the surface area was artificially increased by crushing the shale cores into small particles with diameters in the range of 1.00–1.18 mm (US Mesh 16-18). The shale particles were then collected and stored in a zip-locked bag to avoid oxidation and water uptake.

Pore size distribution of the shale particles was then characterized by N_2 adsorption/desorption test. The N_2 adsorption/desorption test on shale samples #1 and #2 was conducted with the Autosorb iQ-Chemisorption &

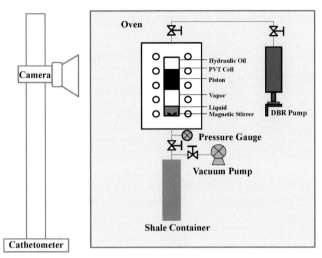

Fig. 2.17 Schematic diagram of the experimental setup for measuring equilibrium P/V isotherms.

Physi-sorption Gas Sorption Analyzer (Quantachrome Instruments, United States). The range of pore width that can be detected by this apparatus is 0.35–500 nm. It took about 3–4 h for N_2 to reach adsorption/desorption equilibrium with the shale samples at a given temperature and pressure.

Furthermore, the pore volume of both shale samples was obtained with the Quantachrome Autosorb software installed for the gas sorption analyzer. Based on the assumption that the pores present in the solid particles are cylindrical, the software can figure out the total pore volume in the particles by measuring the total amount of N_2 take-up at 101.325 kPa and 77 K. From N_2 adsorption/desorption test, the pore volume, and pore size distribution for each shale core were obtained with an accuracy of ±0.6%.

Part of the shale particles were then used for the TOC test. The TOC of both shale cores was measured by a combustion elemental analyzer. During the measurement, H_2SO_4 (10 wt%) was added to the shale particles; then the solution was sparged with oxygen until the purgeable organic carbon and inorganic carbon were removed. The nonpurgeable organic carbon was then placed in the combustion tube to form carbon dioxide, which can be detected by the nondispersive infrared detector. Then, the TOC was obtained. Table 2.2 lists the mass, pore volume, and TOC of the two shale samples.

The system volume was measured after shale core characterization. In order to obtain an accurate P/V isotherm for fluid mixtures in the

Table 2.2 Mineral composition, TOC, pore volume, and the mass of shale sample.

Shale sample	Mass (g)	Pore volume (cm^3/g)	TOC (w/w%)	Clay composition (%)		Other minerals composition (%)			
				Illite	Smectite	Quartz	Potassium feldspar	Plagioclase	Calcite
#1	19.015	0.0211	0.988	31.0	32.0	26.0	1.6	3.0	6.4
#2	18.23	0.014	2.217	25.0	34.0	18.0	3.2	9.7	10.1

partially confined space, it is crucial to obtain an accurate total volume of the mixture. The total volume of the mixture in the partially confined space is calculated as

$$V_{Total} = V_{Cell} + V_{Cell}^{Dead} + V_{Tubing} + V_{Container} + V_{Pore} - V_{Particle} \qquad (2.14)$$

where V_{Total} is the total volume of the mixture in the partially confined space, cm^3; V_{Cell} is the volume of the mixture in the PVT cell, cm^3; V_{Cell}^{Dead} is the dead volume of the PVT cell; V_{Tubing} is the inner volume of stainless steel-tubing lying between the PVT cell and the shale container, cm^3; $V_{Container}$ is the total volume of the shale container, cm^3; V_{Pore} is the total pore volume in the shale particles, cm^3; and $V_{Particle}$ is the total volume of shale particles including both the pore volume and the solid volume, cm^3.

During the experiment, V_{Cell} was obtained by using a cathetometer through measuring the height of the fluid system with an accuracy of $\pm 0.016 \, cm^3$. V_{Cell}^{Dead} is $1.754 \, cm^3$. By employing Boyle-Charles' law, V_{Tubing} was measured as $3.517 \, cm^3$ with an accuracy of $\pm 0.010 \, cm^3$. $V_{Container}$ in this experiment is $10.000 \, cm^3$. V_{Pore} was measured by N_2 adsorption/desorption test. After each P/V isotherm measurement, $V_{Particle}$ was measured through a drainage method by immersing the shale particles into distilled water contained in a cylinder; the volume change before and after the immersion gave the total volume of the shale particles. Considering that the shale material was hydrocarbon-wetting, it is reasonable to assume the distilled water could not enter nanopores in the shale particles. Thus, $V_{Particle}$ is approximately equal to the volume change of water in the cylinder.

Table 2.3 shows the compositions of two $N_2/n\text{-}C_4H_{10}$ mixtures, molar numbers of each component injected and testing temperatures. Before each P/V isotherm measurement, the entire PVT system was tested for leakage with N_2 with testing pressure set as high as $20{,}000 \, kPa$. Then, it was cleaned with acetone, and evacuated using a vacuum pump for 2 h. Shale particles

Table 2.3 Compositions and molar numbers for the N_2/n-C_4H_{10} mixtures tested. For each mixture, experiments were conducted at two temperatures, 299.15 K and 324.15 K.

Molar percentage		Molar numbers	
N_2 (mol%)	n-C_4H_{10} (mol%)	N_2 (mol)	n-C_4H_{10} (mol)
5.40	94.60	0.0140	0.2452
5.01	94.99	0.0131	0.2476

were then placed in the shale container and connected with the PVT cell. The whole system was evacuated again for 12 h using a vacuum pump. In order to prevent shale particles from flowing away during evacuating, a steel mesh with mesh number 200 was used at the outlet of the shale container.

A high-pressure cylinder containing a sufficient amount of n-C_4H_{10} was connected to the PVT cell, allowing direct withdrawal of the liquid n-C_4H_{10} into the PVT cell. A certain amount of liquid n-C_4H_{10} was injected into the PVT cell just above its vapor pressure at room temperature, the air bath temperature was set to be 303.15 K for 12 h, enabling the n-C_4H_{10} sample in the PVT cell to reach thermal equilibrium. Then, the moles of the injected n-C_4H_{10} can be obtained according to its density and its volume measured by the accurate cathetometer. Its density is obtained from the National Institute of Standards and Technology (NIST). Then, N_2 was added into the PVT cell without turning on the magnetic stirrer. The volume of N_2 injected can be determined to be the total volume minus the liquid n-C_4H_{10} volume. No volume change can be assumed since the volume measurement is conducted right after N_2 is injected into the PVT cell. Then, the mass of N_2 added can be determined. Finally, the composition of the binary mixture can be determined. After injection, the mixture was then pressurized into single-liquid state with the magnetic stirrer turned on for 6 h. The mixture was maintained at 303.15 K for 12 h to enable it to reach a thermal equilibrium. Bubble-point pressures for this mixture in the PVT cell were measured by the constant composition expansion (CCE) method. Because the vapor-liquid equilibrium of nonpolar components is predicted quite well with the PR-EOS (Robinson and Peng, 1978; Myers and Sandler, 2002), this EOS was applied to calculate the composition of the mixture. The physical properties of pure components and their binary interaction parameters used with the PR-EOS are shown in Table 2.4. The measured bubble-point pressure was matched by adjusting the overall composition for the PR-EOS model. Then, the matched composition was deemed to be the initial overall composition of the binary mixture; the

Table 2.4 Physical properties of pure components and their binary interaction parameters used in the PR-EOS (1978) model.

Components	Critical temperature (K)	Critical pressure (kPa)	Acentric factor	Binary interaction parameter	
				$n\text{-}C_4H_{10}$	N_2
$n\text{-}C_4H_{10}$	425.2	3799.6875	0.193	0.000	0.095
N_2	126.2	3394.3875	0.040	0.095	0.000

former determined composition of the binary mixture was used to confirm the accuracy of the overall composition determined by the PR–EOS model. This method for determining the composition has an accuracy of ±0.3% on the basis of the crosschecking with the composition measured by gas chromatographic (GC) tests. After the determination of the overall composition, the system temperature was set to an operating temperature. Subsequently, we vigorously stirred the mixture for 6 h by the magnetic stirrer at the selected operating temperature.

P/V isotherms of $N_2/n\text{-}C_4H_{10}$ mixtures were firstly measured in the PVT cell. Subsequently, another set of P/V isotherms for the same mixture were measured in the partially confined space by connecting the PVT cell with the shale container. Crushed shale particles with a certain mass were loaded in the shale container. The container and the tubing between the PVT cell and container were sufficiently vacuumed prior to being connected to the PVT cell. At each temperature, P/V isotherm measurement was initiated from a single-liquid phase state. Then, the pressure was gradually decreased to measure a P/V isotherm for the mixture. The mixture was sufficiently stirred for 30 min to ensure an equilibrium state prior to each volume measurement. After stirring, the magnetic stirrer was switched off and sufficient time, about 4–6 h, was allowed to reach an equilibrium state. The equilibrium was indicated when no pressure changes were observed for a period of 2 h. Thereafter, phase equilibrium of the mixture was visually identified, and the pressure and volume of each phase were measured and recorded. A phase boundary was confirmed by plotting the total volume with respect to pressure and locating the transition point in the curve. The P/V relationship often shows a clear slope change when the vapor phase appears as pressure reduces. The uncertainty in the measurement of the bubble-point pressure as well as the equilibrium pressure and volume is estimated to be ±2.5%. Each measurement is repeated twice to make sure the measured P/V isotherms are reliable and reproducible. The maximum deviation between two consecutive runs is found to be less than ±3.8%.

2.2.2 Characterization of shale samples

Figs. 2.18 and 2.19 show the measured pore size distributions of the shale samples #1 and #2, respectively, as obtained through N_2 adsorption/desorption test. Fig. 2.18 indicates that shale sample #1 contains pores in the range of 1–20 nm. The single sharp peak indicates that shale sample

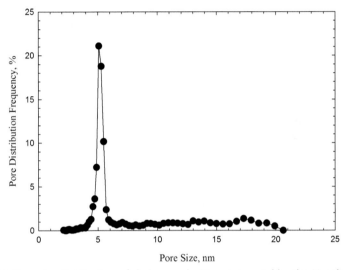

Fig. 2.18 Pore size distribution of shale sample #1 as measured by the N_2 adsorption/desorption test.

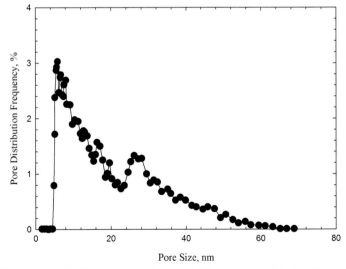

Fig. 2.19 Pore size distribution of shale sample #2 as measured by the N_2 adsorption/desorption test.

#1 has a narrow pore size distribution around 5.0 nm. In contrast, Fig. 2.19 shows that shale sample #2 has a wider pore size distribution in the range of 1–70 nm.

It can be seen from Table 2.2 that the TOC content in shale sample #2 is 2.24 times higher than that in the shale sample #1. Previous studies showed that shale materials with a higher TOC content exhibited a higher sorption capacity (Lu et al., 1995; Jarvie, 2004; Zhang et al., 2012; Clarkson and Haghshenas, 2013). Hence, shale sample #2 is expected to possess a higher sorption capacity than the shale sample #1. In the next section, we will present the sorbed molar numbers of N_2 and n-C_4H_{10} on the two shale samples, and in next Section, we will also explore the possible relationships between TOC content and sorption capacity of individual components.

2.2.3 Phase behavior of N_2/n-C_4H_{10} mixtures in the partially confined space

The P/V isotherms for the N_2/n-C_4H_{10} mixtures, together with the measurement uncertainties, are shown in Figs. 2.20–2.23. In Figs. 2.20–2.23, the solid squares represent the P/V isotherms measured in the PVT cell, and the solid circles represent the P/V isotherms measured in the partially confined space. As depicted in Figs. 2.20–2.23, the phase boundary between single-phase region and two-phase region can be estimated by the intersection of

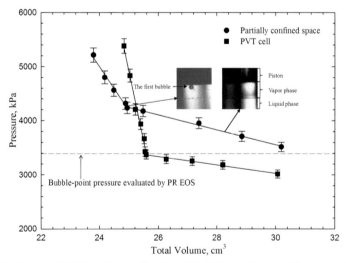

Fig. 2.20 Measured P/V isotherms for the N_2/n-C_4H_{10} mixture with composition of (5.40 mol%, 94.60 mol%) in the PVT cell with and without shale sample #1 at 299.15 K.

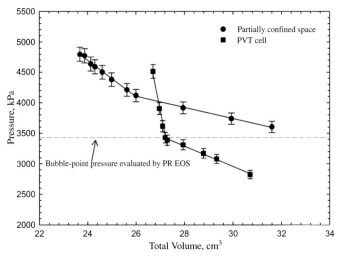

Fig. 2.21 Measured P/V isotherms for the N_2/n-C_4H_{10} mixture with composition of (5.40 mol%, 94.60 mol%) in the PVT cell with and without shale sample #1 at 324.15 K.

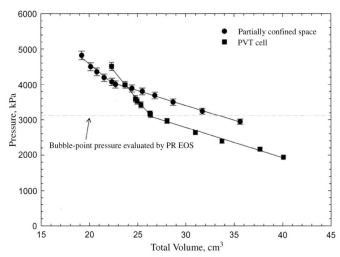

Fig. 2.22 Measured P/V isotherms for the N_2/n-C_4H_{10} mixture with composition of (5.01 mol%, 94.99 mol%) in the PVT cell with and without shale sample #2 at 299.15 K.

two trend lines drawn to represent the two types of phase equilibrium. The dashed line in Figs. 2.20–2.23 represents the bubble-point pressures of the N_2/n-C_4H_{10} mixtures that are calculated with the PR-EOS (1978).

Figs. 2.20–2.23 show that a good agreement is observed between the measured and calculated bubble-point pressures. The bubble-point pressures

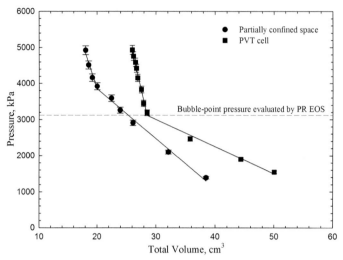

Fig. 2.23 Measured P/V isotherms for the $N_2/n\text{-}C_4H_{10}$ mixture with composition of (5.01 mol%, 94.99 mol%) in the PVT cell with and without shale sample #2 at 324.15 K.

Table 2.5 Changes in the bubble-point pressure of $N_2/n\text{-}C_4H_{10}$ mixtures for two shale samples.

Shale sample	Molar percentage		Temperature (K)	Bubble-point pressure in the PVT cell (kPa)	Bubble-point pressure in the partially confined space (kPa)	Percentage change in bubble-point pressure (%)
	N_2 (mol%)	$n\text{-}C_4H_{10}$ (mol%)				
#1	5.4	94.60	299.15	3390.6	4268.5	25.89
	5.4	94.60	324.15	3429.5	4105.8	19.72
#2	5.01	94.99	299.15	3115.1	4035.5	29.55
	5.01	94.99	324.15	3189.6	3883.8	21.76

Note that the experimental temperatures are above the supercritical temperature of N_2. The bubble-point pressure of $n\text{-}C_4H_{10}$ at 299.15 K and 324.15 K are 241.0 kPa and 523.0 kPa, respectively.

of the $N_2/n\text{-}C_4H_{10}$ mixtures in the partially confined space are higher than those measured in the bulk space. Table 2.5 shows the detailed changes in the bubble-point pressure of the mixtures after being sorbed on the two shale samples. Previous studies reported that when a multicomponent mixture contacts with a shale sample, different components in the mixture exhibit different levels of sorption on shale, leading to the so-called selective sorption phenomenon (Haghshenas et al., 2014; Wang et al., 2015). In this work, the shale particles were always immersed into the single liquid phase during the measurements. The liquid N_2 and n-C_4H_{10} were deemed to sorb on shale samples with different sorption levels at a given temperature and pressure. The selective sorption of N_2 and n-C_4H_{10} changes the initial composition of the N_2/n-C_4H_{10} mixtures, and thus results in an increase in bubble-point pressure, as demonstrated in Figs. 2.20–2.23.

Previous studies demonstrated that bubble-point pressure can be reduced due to the capillary pressure present in small pores (Nojabaei et al., 2013; Devegowda et al., 2012; Teklu et al., 2014). Capillary pressure is attributed to the interfacial tension that exists across the curved interface between the vapor/liquid phases in a tube (Firoozabadi, 2016). Wang et al. (2014, 2016) studied phase behavior of a fluid contained in a porous medium with a pore size distribution; they presented that if the fluid was initially a single-liquid phase during the constant composition expansion (CCE) process, the vapor phase would firstly appear in larger pores, and then appear in smaller pores when all the liquid vaporized in the larger pores. In this study, if we regard the PVT cell as a pore with an infinite pore radius, the first bubble is expected to arise from the PVT cell during the CCE test based on Wang et al. (2014, 2016). During the measurements, the shale samples were always emerged in the single-liquid phase. At the bubble-point, therefore, the fluid contained in the shale pores was always a single-liquid phase and no capillary pressure was present in the shale pores. The measured bubble-point pressure in the partially confined space should be influenced only by the effect of selective sorption between individual components (N_2 and n-C_4H_{10} in this study), without the effect of capillary pressure.

2.2.4 Sorption of individual components on shale samples

This section quantifies the selective sorption between the two components by calculating sorbed molar numbers on the two shale samples based on the results given in Table 2.5. It is explained how selective sorption is expected to affect the bubble-point pressure of a mixture in a partially confined space.

Shale gas-condensate reservoirs, which are normally organic-rich, are traditionally referred as "sorbed gas" reservoirs because a significant amount of shale gas is stored through physical adsorption onto the internal rock surface and through absorption within organic matter (Clarkson and Haghshenas, 2013). Based on the previous research findings, in this research, it can be reasonably deduced that N_2 and n-C_4H_{10} not only adsorb onto the shale rock surface, but also absorb within the organic matter at given temperature and pressure. In addition, Clarkson and Haghshenas (2013) proposed five mechanisms for gas storage in shale gas–condensate reservoirs: (1) Adsorption on internal surface area; (2) Compressed gas storage in natural and hydraulic fractures; (3) Compressed gas storage in matrix porosity; (4) Dissolved gas in formation water; and (5) Absorption in organic matter. Similarly, in our study, N_2 and n-C_4H_{10} in the partially confined space can exhibit two storage states; one is the sorbed gas including adsorption on shale rock surface and absorption in organic matter, and the other is the unsorbed gas which includes the compressed gas located in the pore space inside the shale sample, the bulk space in the PVT cell and in the connecting tubing, and the noncementing pore spaces among the shale particles.

During the P/V isotherm measurements, as the fluid mixture in the partially confined space was depressurized, the first bubble liberated from the single-liquid phase (see Fig. 2.20). At such a bubble point, the composition of the N_2/n-C_4H_{10} mixture in the partially confined space was different from that of the initial mixture loaded in the bulk space. This difference was caused by the selective sorption between N_2 and n-C_4H_{10} on shale samples. In order to obtain the sorbed molar numbers of N_2 and n-C_4H_{10} on the shale samples, we firstly calculated the unsorbed moles of the mixture based on the following three assumptions: (1) The volume of the sorbed layers was negligible in comparison with V_{Total}; (2) The distribution of the unsorbed gas was homogeneous in all spaces; (3) The interactions of unsorbed/sorbed molecules and unsorbed molecules/pore wall were neglected. Then, the following equation was employed to compute the total molar numbers of the unsorbed N_2/n-C_4H_{10} mixture in the partially confined space,

$$n_p = \frac{P_b V_{Total}}{Z_p RT} \qquad (2.15)$$

where n_p is the total molar number of the unsorbed N_2/n-C_4H_{10} mixture in the partially confined space, mol; P_b represents the bubble-point pressure of the unsorbed N_2/n-C_4H_{10} mixture in the partially confined space, Pa; V_{Total} represents the total volume of the unsorbed N_2/n-C_4H_{10} mixture in the partially confined space, m^3, Z_p is the compressibility factor of the unsorbed N_2/n-C_4H_{10} mixture in the partially confined space that is calculated by the PR–EOS; R is the universal gas constant, $8.314 \, m^3 \, Pa \, K^{-1} \, mol^{-1}$; and T is the temperature, K.

After obtaining the total molar numbers of the unsorbed N_2/n-C_4H_{10} mixture in the partially confined space, the sorbed molar numbers of N_2 and n-C_4H_{10} can then be determined at the bubble point with the following equations, respectively,

$$n_{ad\_N_2} = n_o s_{N_2} - n_p x'_{N_2} \qquad (2.16)$$

$$n_{ad\_C_4} = n_o x_{C_4} - n_p x'_{C_4} \qquad (2.17)$$

where $n_{ad\_N_2}$ and $n_{ad\_C_4}$ are the sorbed molar numbers of N_2 and n-C_4H_{10} on shale samples, respectively, mol; n_o is the injected molar number of the N_2/n-C_4H_{10} mixture, mol; x_{N_2} and x_{C_4} are the molar percentage of N_2 and n-C_4H_{10} in the injected mixture in the PVT cell, respectively; x'_{N_2} and x'_{C_4} are the molar percentage of N_2 and n-C_4H_{10} when the mixture rests at the bubble-point in the partially confined space, respectively. Eq. (2.16) and (2.17) are derived based on the assumption that the distribution of the unsorbed gas is uniform in all spaces as mentioned above.

Table 2.5 lists the calculated molar percentages of N_2 and n-C_4H_{10} in the mixture before and after sorption, and sorbed molar numbers of N_2 and n-C_4H_{10} in the partially confined spaces. The higher sorption tendency of n-C_4H_{10} than that of N_2 is indicated in calculated sorbed molar numbers in Table 2.5. The molar percentage of N_2 in mixture in the partially confined space tends to increase in comparison to that in the bulk space. The higher N_2 concentration results in the higher bubble-point pressure of the N_2/n-C_4H_{10} mixture in the partially confined space.

Table 2.5 also shows that, for a given N_2/n-C_4H_{10} mixture, both N_2 and n-C_4H_{10} sorb more at a lower temperature. A higher temperature did not lead to a higher bubble-point pressure for a given mixture (Table 2.4) likely because bubble-point is more sensitive to composition than to temperature for these mixtures at the conditions tested. This emphasizes the importance of considering sorption in phase behavior calculation for small pores.

2.2.5 Effect of TOC on sorption capacity

Sorption capacity is defined as the ability of gas storage on shale; quantitatively, it is equal to the sorbed molar numbers per gram of shale rocks. Organic matters present in shale rocks, generally represented by TOC content, can sorb and store shale components. Previous studies have investigated the relationship between TOC content and sorption capacity, showing that an increase in TOC content can lead to an approximately linear increase in the sorption capacity (Lu et al., 1995; Jarvie, 2004; Zhang et al., 2012). Table 2.6 also shows, given the fact that the two N_2/n-C_4H_{10} mixtures have a similar composition, much more N_2 and n-C_4H_{10} are sorbed onto the shale sample #2 than the shale sample #1 under similar temperature/pressure conditions. This can be attributed to a higher TOC content in the shale sample #2 than the shale sample #1.

Table 2.6 Calculated molar percentages of N$_2$ and n-C$_4$H$_{10}$ in the mixture before and after sorption, and sorbed molar numbers of N$_2$ and n-C$_4$H$_{10}$ at the bubble-point in the partially confined spaces.

Shale sample	Temperature (K)	Molar percentage before sorption		Molar percentage after sorption		Adsorbed molar numbers	
		N$_2$ (mol%)	n-C$_4$H$_{10}$ (mol%)	N$_2$ (mol%)	n-C$_4$H$_{10}$ (mol%)	N$_2$ (mol)	n-C$_4$H$_{10}$ (mol)
#1	299.15	5.40	94.60	7.00	93.00	0.0007	0.0683
#1	324.15	5.40	94.60	6.70	93.30	0.0006	0.0583
#2	299.15	5.01	94.99	6.59	93.41	0.0008	0.0728
#2	324.15	5.01	94.99	6.25	93.75	0.0007	0.0612

Note that the molar concentrations of N$_2$ and n-C$_4$H$_{10}$ have been determined by the PR-EOS (1978) calibrated with the measured bubble-point.

Sorption capacities of N$_2$ and n-C$_4$H$_{10}$ are closely correlated with the TOC content in the shale samples. Charoensuppanimit et al. (2016) measured N$_2$ and CH$_4$ sorption on shale materials and found both of their sorption capacities are positively correlated with TOC content. As for the sorption of n-C$_4$H$_{10}$ on shale materials, however, there are no data available in the literature. Considering that CH$_4$ and n-C$_4$H$_{10}$ are both hydrocarbons and have natural affinity with TOC content, it is conceivable that the sorption capacity for n-C$_4$H$_{10}$ is correlated with TOC content positively.

After obtaining the sorbed molar numbers of individual components in each shale sample, the sorption capacities of N$_2$ and n-C$_4$H$_{10}$ on two shale samples can be calculated by the following equation,

$$V_{ad} = \frac{n_{ad}}{m} \tag{2.18}$$

where V_{ad} represents the sorption capacity, mmol/g; n_{ad} is the sorbed molar numbers of individual components in each shale sample, mmol; m is the mass of the shale sample, g.

Figs. 2.24 and 2.25, respectively, show the sorption capacities of N$_2$ and n-C$_4$H$_{10}$ in terms of the TOC content in both shale samples. The sorption capacity for n-C$_4$H$_{10}$ increases with the increasing TOC content, while the sorption capacity for N$_2$ only increases slightly with the TOC content. A shale sample with a higher TOC content is expected to have a higher sorption capacity, which is consistent with the findings by Nuttall et al. (2005) that sorption occurs primarily on active sites containing organic carbons. Besides, the natural affinity between hydrocarbon n-C$_4$H$_{10}$ and TOC content leads to a higher sorption of n-C$_4$H$_{10}$ as compared to nonhydrocarbon

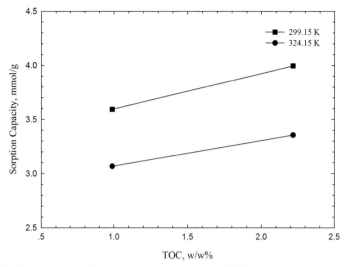

Fig. 2.24 Sorption capacity of n-C_4H_{10} in terms of TOC content on the two shale samples.

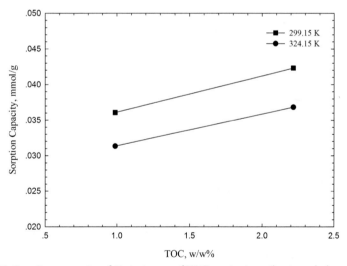

Fig. 2.25 Sorption capacity of N_2 in terms of TOC content on the two shale samples.

N_2. However, it seems that the sorption amount is not affected much by the TOC content of shale sample #1 and #2. Recently, Xiong et al. (2017) conducted a series of methane sorption measurements for seven shale core samples collected from the Ordos Basin with depths over 3000 m and TOCs ranging from 0.49% to 3.82%. They then proposed that the sorption capacity did not correlate only with the TOC content but showed a more complex dependence on petrophysical and mineralogical properties. Therefore, to understand the sorption capacity of N$_2$ or n-C$_4$H$_{10}$, more sorption data should be measured and other factors, e.g., clay minerals should be considered to have a better understanding on the sorption behavior.

References

Alfi, M., Nasrabadi, H., Banerjee, D., 2016. Experimental investigation of confinement effect on phase behavior of hexane, heptane and octane using lab-on-a-chip technology. Fluid Phase Equilib. 423, 25–33.

Alharthy, N.S., Nguyen, T., Teklu, T., Kazemi, H., Graves, R., 2013. Multiphase compositional modeling in small-scale pores of unconventional shale reservoirs. In: SPE Annual Technical Conference and Exhibition. New Orleans, Louisiana, Society of Petroleum Engineers.

Balbuena, P.B., Gubbins, K.E., 1993. Theoretical interpretation of adsorption behavior of simple fluids in slit pores. Langmuir 9 (7), 1801–1814.

Basaldella, E.I., Tara, J.C., Armenta, G.A., Iglesias, M.E.P., 2007. Cu/SBA-15 as adsorbent for propane/propylene separation. J. Porous. Mater. 14 (3), 273–278.

Bruot, N., Caupin, F., 2016. Curvature dependence of the liquid-vapor surface tension beyond the Tolman approximation. Phys. Rev. Lett. 116 (5), 056102.

Bui, K., Akkutlu, I.Y., 2015. Nanopore wall effect on surface tension of methane. Mol. Phys. 113 (22), 3506–3513.

Cabral, V.F., Alfradique, M.F., Tavares, F.W., Castier, M., 2005. Thermodynamic equilibrium of adsorbed phases. Fluid Phase Equilib. 233 (1), 66–72.

Cervilla, A., Corma, A., Fornes, V., Llopis, E., Palanca, P., Rey, F., Ribera, A., 1994. Intercalation of [MoVIO$_2$(O$_2$CC(S)Ph$_2$)2]$_2$-in a Zn (II)-Al (III) layered double hydroxide host: a strategy for the heterogeneous catalysis of the air oxidation of thiols. J. Am. Chem. Soc. 116 (4), 1595–1596.

Charoensuppanimit, P., Mohammad, S.A., Gasem, K.A.M., 2016. Measurements and modeling of gas adsorption on shales. Energy Fuel 30, 2309–2319.

Civan, F., Devegowda, D., Sigal, R., Vafai, K., 2012. Theoretical fundamentals, critical issues, and adequate formulation of effective shale gas and condensate reservoir simulation. AIP Conf. Proc. 1453 (1), 155–160.

Clarkson, C.R., Haghshenas, B., 2013. Modeling of supercritical fluid adsorption on organic-rich shales and coal. In: SPE Unconv. Resour. Conf. Woodlands, Texas, USA.

de Boer, J.H., Lippens, B.C., 1964. Studies on pore systems in catalysts II. The shapes of pores in aluminum oxide systems. J. Catal. 3 (1), 38–43.

Devegowda, D., Sapmanee, K., Civan, F., Sigal, R., 2012. Phase behavior of gas condensates in shale due to pore proximity effects: implications for transport, reserves and well productivity. In: SPE Annu. Tech. Conf. Exhib., San Antonio, Texas, USA.

Didar, B.R., Akkutlu, I.Y., 2013. Pore-size dependence of fluid phase behavior and properties in organic-rich shale reservoirs. In: SPE International Symposium on Oilfield Chemistry. The Woodlands, Texas, USA Society of Petroleum Engineers.

Dong, X., Liu, H., Hou, J., Wu, K., Chen, Z., 2016. Phase equilibria of confined fluids in nanopores of tight and shale rocks considering the effect of capillary pressure and adsorption film. Ind. Eng. Chem. Res. 55, 798–811.

Ebner, C., Saam, W.F., 1977. New phase-transition phenomena in thin argon films. Phys. Rev. Lett. 38 (25), 1486–1489.

Ebner, C., Saam, W.F., Stroud, D., 1976. Density-functional theory of simple classical fluids. I. Surfaces. Phys. Rev. A 14 (6), 2264–2273.

Fan, C., Do, D.D., Nicholson, D., 2011. On the cavitation and pore blocking in slit-shaped ink-bottle pores. Langmuir 27 (7), 3511–3526.

Firoozabadi, A., 2016. Thermodynamics and Applications in Hydrocarbon Energy Production. McGraw Hill, New York.

Gasparik, M., Ghanizadeh, A., Gensterblum, Y., Krooss, B.M., Littke, R., 2012. The methane storage capacity of black shales. In: 3rd EAGE Shale Wrokshop.

Haghshenas, B., Soroush, M., Broh, I.I., Clarkson, C.R., 2014. Simulation of liquid-rich shale gas reservoirs with heavy hydrocarbon fraction desorption. In: SPE Unconv. Resour. Conf. Woodlands, Texas, USA.

Jarvie, D., 2004. Evaluation of Hydrocarbon Generation and Storage in Barnett Shale, Fort Worth Basin. The University of Texas at Austin, Bureau of Economic logy/PTTC, Texas.

Jhaveri, B.S., Youngren, G.K., 1988. Three-parameter modification of the peng-robinson equation of state to improve volumetric predictions. SPE Reserv. Eng. 3 (03), 1033–1040.

Jin, Z., 2017. Bubble/dew point and hysteresis of hydrocarbons in nanopores from molecular perspective. Fluid Phase Equilib. 458, 177–185.

Jin, Z., Firoozabadi, A., 2016a. Phase behavior and flow in shale nanopores from molecular simulations. Fluid Phase Equilib. 430, 156–168.

Jin, Z., Firoozabadi, A., 2016b. Thermodynamic modeling of phase behavior in shale media. SPE J. 21 (1), 190–207.

Jin, B., Nasrabadi, H., 2016. Phase behavior of multi-component hydrocarbon systems in nano-pores using gauge-GCMC molecular simulation. Fluid Phase Equilib. 425, 324–334.

Jin, L., Ma, Y., Jamili, A., 2013. Investigating the effect of pore proximity on phase behavior and fluid properties in shale formations. In: SPE Annual Technical Conference and Exhibition. New Orleans, Louisiana, Society of Petroleum Engineers.

Jin, B., Bi, R., Nasrabadi, H., 2017. Molecular simulation of the pore size distribution effect on phase behavior of methane confined in nanopores. Fluid Phase Equilib. 452 (Suppl. C), 94–102.

Klomkliang, N., Do, D.D., Nicholson, D., 2013. On the hysteresis loop and equilibrium transition in slit-shaped ink-bottle pores. Adsorption 19 (6), 1273–1290.

Lev, D.G., Gubbins, K.E., Radhakrishnan, R., Sliwinska-Bartkowiak, M., 1999. Phase separation in confined systems. Rep. Prog. Phys. 62 (12), 1573.

Li, Z., Firoozabadi, A., 2009. Interfacial tension of nonassociating pure substances and binary mixtures by density functional theory combined with Peng–Robinson equation of state. J. Chem. Phys. 130 (15), 154108.

Li, Z., Jin, Z., Firoozabadi, A., 2014. Phase behavior and adsorption of pure substances and mixtures and characterization in nanopore structures by density functional theory. SPE J. 19 (6), 1096–1109.

Lu, X., Li, F., Watson, A.T., 1995. Adsorption measurements in Devonian shale. Fuel 74, 599–603.

Luo, S., Lutkenhaus, J.L., Nasrabadi, H., 2016a. Confinement-induced supercriticality and phase equilibria of hydrocarbons in nanopores. Langmuir 32 (44), 11506–11513.

Luo, S., Lutkenhaus, J.L., Nasrabadi, H., 2016b. Use of differential scanning calorimetry to study phase behavior of hydrocarbon mixtures in nano-scale porous media. J. Pet. Sci. Eng. 163, 731–738.

Luo, S., Nasrabadi, H., Lutkenhaus, J.L., 2016c. Effect of confinement on the bubble points of hydrocarbons in nanoporous media. AICHE J. 62 (5), 1772–1780.

Luo, S., Lutkenhaus, J.L., Nasrabadi, H., 2017. Multi-scale fluid phase behavior simulation in shale reservoirs by a pore-size-dependent equation of state. In: SPE Annual Technical Conference and Exhibition. San Antonio, USA, Society of Petroleum Engineers.

Morishige, K., Fujii, H., Uga, M., Kinukawa, D., 1997. Capillary critical point of argon, nitrogen, oxygen, ethylene, and carbon dioxide in MCM-41. Langmuir 13, 3494–3498.

Myers, J.A., Sandler, S.I., 2002. An equation of state for electrolyte solutions covering wide ranges of temperature, pressure, and composition. Ind. Eng. Chem. Res. 41, 3282–3297.

Nagarajan, N.R., Hanapour, M.M., Arasteh, F., 2013. Critical role of rock and fluid impact on reservoir performance on unconventional shale reservoirs. In: Unconv. Resour. Tech. Conf. Denver, Colorado, USA.

Neimark, A.V., Vishnyakov, A., 2000. Gauge cell method for simulation studies of phase transitions in confined systems. Phys. Rev. E 62 (4), 4611–4622.

NIST, 2017. Chemistry WebBook, NIST Standard Reference Database Number 69. National Institute of Standards and Technology, Gaithersburg, MD. http://webbook. nist.gov.

Nojabaei, B., Johns, R.T., Chu, L., 2013. Effect of capillary pressure on phase behavior in tight rocks and shales. SPE Reserv. Eval. Eng. 16, 281–289.

Nuttall, B.C., Eble, C.F., Drahovzal, J.A., Bustin, M., 2005. Analysis of Devonian Black Shales in Kentucky for Potential Carbon Dioxide Sequestration and Enhanced Natural Gas Production. University of Kentucky.

Parsa, E., Yin, X., Ozkan, E., 2015. Direct observation of the impact of nanopore confinement on petroleum gas condensation. In: SPE Annual Technical Conference and Exhibition. Houston, USA, Society of Petroleum Engineers.

Peng, D.-Y., Robinson, D.B., 1976. A new two-constant equation of state. Ind. Eng. Chem. Fundam. 15 (1), 59–64.

Robinson, D.B., Peng, D.Y., 1978. The Characterization of the Heptanes and Heavier Fractions for the GPA Peng-Robinson Programs, Gas Processors Association. Research Report RR-28. (Booklet only sold by the Gas Processors Association, GPA).

Robinson, D.B., Peng, D.-Y., Chung, S.Y.K., 1985. The development of the Peng-Robinson equation and its application to phase equilibrium in a system containing methanol. Fluid Phase Equilib. 24 (1–2), 25–41.

Rosenfeld, Y., 1989. Free-energy model for the inhomogeneous hard-sphere fluid mixture and density-functional theory of freezing. Phys. Rev. Lett. 63 (9), 980–983.

Rowlinson, J.S., Swinton, F.L., 1982. Liquids and Liquid Mixtures. Butterworth, London.

Rowlinson, J.S., Widom, B., 1982. Molecular Theory of Capillarity. Clarendon, Oxford.

Sandoval, D.R., Yan, W., Michelsen, M.L., Stenby, E.H., 2016. The phase envelope of multicomponent mixtures in the presence of a capillary pressure difference. Ind. Eng. Chem. Res. 55 (22), 6530–6538.

Santiso, E., Firoozabadi, A., 2006. Curvature dependency of surface tension in multicomponent systems. AICHE J. 52 (1), 311–322.

Sapmanee, K., 2011. Effects of Pore Proximity on Behavior and Production Prediction of Gas/Condensate. University of Oklahoma.

Sing, K.S.W., Everett, D.H., Haul, R.A.W., Moscou, L., Pierotti, R.A., Rouquerol, J., Siemieniewska, T., 2008. Reporting physisorption data for gas/solid systems. In: Handbook of Heterogeneous Catalysis. Wiley-VCH Verlag GmbH & Co. KGaA.

Singh, J.K., Kwak, S.K., 2007. Surface tension and vapor-liquid phase coexistence of confined square-well fluid. J. Chem. Phys. 126 (2), 024702.

Singh, S.K., Sinha, A., Deo, G., Singh, J.K., 2009. Vapor–liquid phase coexistence, critical properties, and surface tension of confined alkanes. J. Phys. Chem. C 113 (17), 7170–7180.

Steele, W.A., 1973. The physical interaction of gases with crystalline solids: I. Gas-solid energies and properties of isolated adsorbed atoms. Surf. Sci. 36 (1), 317–352.

Tan, S.P., Piri, M., 2015. Equation-of-state modeling of confined-fluid phase equilibria in nanopores. Fluid Phase Equilib. 393, 48–63.

Teklu, T.W., Alharthy, N., Kazemi, H., Yin, X., Graves, R.M., Alsumaiti, A.M., 2014. Phase behavior and minimum miscibility pressure in nanopores. SPE Reserv. Eval. Eng. 17, 396–403.

Travalloni, L., Castier, M., Tavares, F.W., Sandler, S.I., 2010a. Critical behavior of pure confined fluids from an extension of the van der Waals equation of state. J. Supercrit. Fluids 55 (2), 455–461.

Travalloni, L., Castier, M., Tavares, F.W., Sandler, S.I., 2010b. Thermodynamic modeling of confined fluids using an extension of the generalized van der Waals theory. Chem. Eng. Sci. 65 (10), 3088–3099.

Travalloni, L., Castier, M., Tavares, F.W., 2014. Phase equilibrium of fluids confined in pores media from an extended Peng-Robinson equation of state. Fluid Phase Equilib. 362, 335–341.

Volzone, C., 2007. Retention of pollutant gases: comparison between clay minerals and their modified products. Appl. Clay Sci. 36 (1–3), 191–196.

Walton, J.P.R.B., Quirke, N., 1989. Capillary condensation: a molecular simulation study. Mol. Simul. 2 (4-6), 361–391.

Wang, L., Parsa, E., Gao, Y.F., Ok, J.T., Neeves, K., Yin, X.L., Ozkan, E., 2014. Experimental study and modeling of the effect of nanoconfinement on hydrocarbon phase behavior in unconventional reservoirs. In: SPE Western North American and Rocky and Rocky Mountain Joint Regional Meeting, Denver, Colorado, USA.

Wang, Y., Tsotsis, T.T., Jessen, K., 2015. Competitive sorption of methane/ethane mixtures on shale: measurements and modeling. Ind. Eng. Chem. Res. 54, 12178–12195.

Wang, L., Yin, X., Neeves, K.B., Ozkan, E., 2016. Effect of pore-size distribution on phase transition of hydrocarbon mixtures in nanoporous media. SPE J. 21 (6), 1981–1995.

Weinaug, C.F., Katz, D.L., 1943. Surface tensions of methane-propane mixtures. Ind. Eng. Chem. 35 (2), 239–246.

Wongkoblap, A., Do, D.D., Birkett, G., Nicholson, D., 2011. A critical assessment of capillary condensation and evaporation equations: a computer simulation study. J. Colloid Interface Sci. 356 (2), 672–680.

Xiong, F., Wang, X., Amooie, N., et al., 2017. The shale gas sorption capacity of transitional shales in the Ordos Basin, NW China. Fuel 208, 236–246.

Yan, X., Wang, T.B., Gao, C.F., Lan, X.Z., 2013. Mesoscopic phase behavior of tridecane-tetradecane mixtures confined in porous materials: effects of pore size and pore geometry. J. Phys. Chem. C 117, 17245–17255.

Zhang, T., Ellis, G.S., Ruppel, S.C., Milliken, K., Yang, R., 2012. Effect of organic-matter type and thermal maturity on methane adsorption in shale-gas systems. J. Org. Geochem. 47, 120–131.

Zhang, Y., Civan, F., Devegowda, D., Sigal, R.F., 2013. Improved prediction of multicomponent hydrocarbon fluid properties in organic rich shale reservoirs. In: SPE Annual Technical Conference and Exhibition. New Orleans, Louisiana.

Zhong, J., Zandavi, S.H., Li, H., Bao, B., Persad, A.H., Mostowfi, F., Sinton, D., 2017. Condensation in one-dimensional dead-end nanochannels. ACS Nano 11 (1), 304–313.

CHAPTER THREE

Adsorption behavior of reservoir fluids and CO_2 in shale

Shale oil/gas resources, as an important unconventional hydrocarbon resource, have been attracting global attention in recent years due to their considerable abundance (Huang et al., 2018a,b; Yuan et al., 2015a,b; Weijermars, 2014; Yamazaki et al., 2006; Karacan et al., 2011). However, the unique characteristics of shale reservoirs, such as extremely low permeability and heterogeneity, make it difficult to recover shale resources from such reservoirs (Weijermars, 2013). Unlike conventional reservoirs, pores in shale matrix are generally in the nanometer range. In nanopores, the fluid-pore surface interactions are significant, leading to strong adsorption of components on the pore surface. Moreover, individual components generally exhibit different levels of adsorption capacity on the pore surface, resulting in the competitive adsorption phenomenon (Wang et al., 2015). Understanding of the competitive adsorption behavior of fluids in shale is of critical importance for more accurately determining the macroscopic and microscopic distribution of fluids in shale reservoirs as well as the mechanisms governing the fluid transport in shale reservoirs.

A number of theoretical methods have been employed to study the adsorption behavior of fluid on shale. Langmuir adsorption model and Brunauer-Emmett-Teller (BET) method were initially adopted to describe the adsorption behavior in shale media (Ambrose et al., 2010; Weniger et al., 2010; Brunauer et al., 1938; Langmuir, 1916; Halsey, 1948; Roque-Malherbe, 2007). Although the Langmuir model with tuned parameters can give a satisfying match with experimental data, this model is derived from the assumption that the fluid molecules form single-adsorption layer on pore surface, while the remaining molecules distribute in pores homogeneously (Li et al., 2014). Moreover, the single adsorption-layer assumption may not be valid in nanopores. It has been found that molecules can exhibit multilayered adsorption on pore surface due to the strong interactions between molecules and pore surface (Li et al., 2014). The BET model is proposed under the assumption that the fluid molecules can form an infinite number of adsorption layers, while it still assumes homogeneous fluid distributions

Confined Fluid Phase Behavior and CO_2
Sequestration in Shale Reservoirs
https://doi.org/10.1016/B978-0-323-91660-8.00009-9

by neglecting the fluid-pore surface interactions (Li et al., 2014). The ideal adsorbed solution theory (IAST) was proposed to estimate the mixture adsorption in porous media (Myers and Prausnitz, 1965; Cessford et al., 2012), with which the multicomponent adsorption isotherms can be derived solely from single-component adsorption data. However, this method only works well for ideal systems (Sweatman and Quirke, 2002).

The grand canonical Monte Carlo (GCMC) simulations (Jin and Nasrabadi, 2016; Neimark and Vishnyakov, 2000; Singh et al., 2009; Walton and Quirke, 1989; Wongkoblap, et al., 2011), density functional theory (DFT) (Liu et al., 2018a,b; Jin and Firoozabadi, 2016a,b; Li, et al., 2014), and molecular dynamics (MD) simulations (Jin and Firoozabadi, 2015; Shirono and Daiguiji, 2007) have been recently adopted to study the adsorption behavior of fluids in nanoscale pores. These statistical thermodynamic approaches can explicitly consider the fluid-pore surface interactions and have shown excellent agreement with the experimental data on gas adsorption and interfacial phenomena (Liu et al., 2017; Li and Firoozabadi, 2009). On the basis of the calculations made with molecular simulation methods, density distribution of molecules in nanopores is heterogeneous: near the pore surface, the density is significantly higher than that at the pore center, while the density in pore center approaches the bulk (Tian et al., 2017). But these existing molecule-based models use a single pore size to describe the adsorption behavior. However, in shale reservoirs, the sizes of nanopores are generally not uniform, exhibiting different pore-size distributions in different parts of the reservoir. Thereof, to better understand the adsorption of fluids in nanopores, the effect of pore-size distribution should be considered.

Injection of CO_2 into shale reservoirs has been proposed as a promising method that can not only enhance shale oil/gas recovery but also sequestrate CO_2 in shale reservoirs (White et al., 2005; Chatterjee and Paul, 2013; Vishal et al., 2015; Jiang, 2011; Huang et al., 2019). Since C_1 is generally the dominant component in shale fluids, studies regarding enhancing C_1 recovery with CO_2 are mostly conducted. Recently, experimental measurements at the laboratory scale are conducted to investigate the adsorption behavior of C_1/CO_2 mixtures on shale rocks (Gensterblum et al., 2014; Ottiger et al., 2008; Bhowmik and Dutta, 2011; Faiz et al., 2007; Khosrokhavar et al., 2014; Busch et al., 2003; Majewska et al., 2009; Ross and Bustin, 2007). It is found that CO_2 has a higher adsorption capacity than C_1, indicating that CO_2 injection can be an effective method for enhancing shale C_1 recovery in shale reservoirs. Numerical simulations have also been conducted to validate the feasibility of this technique (Kim et al., 2017a,b; Luo et al., 2013; Yu et al., 2014; Jiang et al., 2014; Godec et al., 2013; Liu et al., 2013, 2016; Zhang et al., 2015; Kazemi and Takbiri-Borujeni, 2016; Kurniawan et al., 2006; Yuan et al., 2015a,b; Brochard et al., 2012; Wang et al., 2016a,b; Lu et al., 2015; Sun et al., 2017; Huang et al., 2018a,b; Wu et al., 2015a,b; Kowalczyk et al., 2012). Numerical studies indicated that CO_2 injection into depleted shale gas reservoirs for enhanced gas recovery is technically feasible. Although these studies can provide us some insights on the competitive adsorption behaviors of C_1/CO_2 mixtures, the effect of CO_2 on the recovery of heavier hydrocarbons is scarcely studied. Recently, Jin and Firoozabadi (2016a,b) investigated the competitive adsorption of nC_4/CO_2 mixture in organic pores, finding that using CO_2 to recover nC_4 from nanopores is more significant at higher pressures. However, again, this study only investigates adsorption behavior in single nanopore without the consideration of pore-size distribution.

The study on adsorption behavior of hydrocarbon(s)/CO_2 mixtures in shale is still at a preliminary stage. In this section, a double-nanopore system is built comprising of two pores with different pore sizes and investigates the competitive adsorption behavior of hydrocarbon(s)/CO_2 mixtures in this double-nanopore system using the MD method. The adsorption behavior of hydrocarbon(s)/CO_2 mixtures is studied with a depressurization method.

3.1 Competitive adsorption behavior of hydrocarbons and hydrocarbon/CO_2 mixtures in porous media from molecular perspective

3.1.1 Methodology and simulation model

3.1.1.1 Molecular dynamic simulation

The Forcite Module and Amorphous Cell Package in Material Studio software is employed (Yu et al., 2017; Zhao et al., 2016; Valentini et al., 2011; Li et al., 2016). Within the MD simulations, the condensed-phased-optimized molecular potential for atomistic simulation studies (COMPASS) force field is applied to describe the interatomic interactions in the Force Module. The COMPASS force-field was originally proposed by Sun (1998). It has been recognized as the first force-field that enables an accurate simultaneous prediction for a broad range of molecules, including polymers (Li et al., 2016; Sun, 1998). More details about the COMPASS force-field terms and parameters can be found in Sun (1998). The COMPASS force-field is only explained concisely in this work. In the COMPASS force field, the total potential energy (E^{total}) is given as (Rigby et al., 1997):

$$E^{total} = E^{internal} + E^{cross-coupling} + E^{vanderWaals} + E^{electrostatics} \tag{3.1}$$

$$E^{internal} = \sum E^{(b)} + \sum E^{(\theta)} + \sum E^{(\varphi)} + \sum E^{(\gamma)} \tag{3.2}$$

$$E^{cross-coupling} = \sum E^{(b\theta)} + \sum E^{(b\varphi)} + \sum E^{(b'\varphi)} + \sum E^{(\theta\theta')}$$
$$+ \sum E^{(\theta\varphi)} + \sum E^{(\theta\theta'\varphi)} \tag{3.3}$$

where b and b' represent the lengths of two adjacent bonds, respectively; θ and θ' represent the angles between two adjacent bonds, respectively; ϕ is the angle resulted from dihedral torsion; and γ represents the out of the plane angle (Li et al., 2016). $E^{internal}$ is the energy from each of the internal valence coordinates; $E^{cross-coupling}$ is the cross–coupling term between internal coordinates. $E^{vanderWaals}$ is represented by a sum of repulsive and attractive Lennard-Jones terms (Jones, 1924). In the molecular model, the vdW interactions are calculated using the atom-based method with a cut-off distance of 12.5 Å, while the Ewald method is employed to calculate the electrostatic interactions with an accuracy of 1.0×10^{-4} kcal/mol (Song et al., 2017). Additionally, the Andersen thermostat (Andersen, 1980) is used for temperature conversion.

3.1.1.2 Simulation model

Carbon materials have been widely used to simulate kerogen in shale considering that kerogen is hydrophobic (Li et al., 2014; Liu et al., 2018a,b). The full atomistic structure of graphite layers is used, formed by carbon atoms, to simulate the organic pores. As shown in Fig. 3.1, two carbon–slit

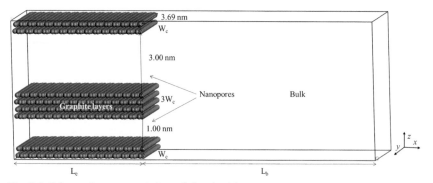

Fig. 3.1 Schematic representation of the double-pore system.

pores are placed in parallel and are connected by a space with given volume. The graphene layers located in the left-hand of the pore system create a confined space (which can represent organic pores in shale), while the space in the right hand of the pore system corresponds to a bulk space (which can represent a microfracture). Overall, this double-nanopore system consists of three regions, i.e., the 1-nm pore, the 3-nm pore, and a bulk space connecting the two pores. As for this pore system, the periodic boundary condition is applied in all three directions. In this simulation box, four graphic layers are used to form one carbon sheet and three carbon sheets are used to construct the two carbon-slit pores. The separation between two carbon-atom centers residing in the two graphite layers is 0.335 nm. Fig. 3.2 presents the schematic of the carbon sheet. As shown in Fig. 3.2, the distance between two adjacent carbon-atom centers in the same graphite layer is 0.142 nm. During the simulation, the position of carbon sheets is fixed. The size of the simulation box is $(L_c + L_b)$ nm $\times$ 3.69 nm $\times (4 + 5W_c)$ nm in the x, y, and z directions, respectively (L_c is the length of the two pores, 6.619 nm; L_b is the length of the connecting space; and W_c is the separation distance between the two carbon-atom centers in the two graphite layers, 0.335 nm).

MD simulations are performed in the canonical NVT ensemble; such an ensemble has fixed number of particles (N), volume (V), and temperature (T). It is initially loaded with a mixture composed of given number of molecules in the NVT ensemble. The pressure of the system is decreased by increasing the length of the bulk space. Three pressure conditions (i.e., 3.97, 5.66, and 7.94 MPa) are created by changing L_b to be 3.322 nm, 11.381 nm, and 26.381 nm, respectively. Based on the known system volume, system temperature, and mole numbers of the molecules in the system, the system pressure is calculated using the Peng-Robinson equation of state (1978). The basic inputs used for MD simulations are summarized in Table 3.1. It should be mentioned that the COMPASS force field has been

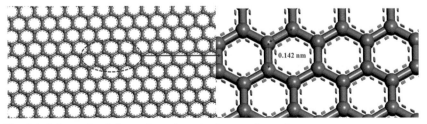

Fig. 3.2 Schematic of the carbon sheet in the x-y plane view.

Table 3.1 Summary of the parameter values used in the MD simulations.

Mixtures	Molar fraction	Simulation pressure (MPa)	Simulation time (ns)	Simulation temperature (K)
C_1/nC_4	0.7:0.3	3.97, 5.66, 7.94	35	333.15
C_1/CO_2	0.5:0.5	3.97, 5.66, 7.94	35	333.15
nC_4/CO_2	0.5:0.5	3.97, 5.66, 7.94	35	333.15
$nC_4/C_1/CO_2$	0.15:0.35:0.5	3.97, 5.66, 7.94	35	333.15

widely used for hydrocarbons for hydrocarbons and CO_2 (Rao et al., 2012; Ta et al., 2015), validating its efficiency in describing C_1, nC_4, and CO_2 in our model.

When placed in this pore system, C_1, nC_4, and CO_2molecules in the C_1/nC_4, C_1/CO_2, and nC_4/CO_2 mixtures generally display competitive adsorption on pore surface at given temperature and pressure conditions. Selectivity coefficient is used to characterize the competitive adsorption of C_1, nC_4, and CO_2 molecules in organic pores (Kurniawan et al., 2006). As for binary mixtures, the selectivity coefficient ($S_{A/B}$) is defined as (Kurniawan et al., 2006),

$$S_{A/B} = \frac{(x_A/x_B)}{(\gamma_A/\gamma_B)} \qquad (3.4)$$

where A and B are the species in binary mixtures; x_A and x_B are molar fractions of adsorbates A and B in the adsorbed phase, respectively; γ_A and γ_B are the molar fractions of adsorbates A and B in the space connecting two pores, respectively. If $S_{A/B}$ is less than 1, it indicates that the adsorption capacity of A is lower than that of B; that is, B is easier to adsorb on pore surface than A (Liu and Wilcox, 2012).

3.1.2 Adsorption behavior of mixtures in the double-nanopore system

(1) **Competitive adsorption behavior of C_1/nC_4 mixture**. Fig. 3.3 presents the density profiles of C_1/nC_4 mixture (70 mol%:30 mol%) in the two pores at $T=333.15$ K and three different pressures. Near the pore wall, the in situ density of C_1 and nC_4 molecules is much higher than that at the pore center, indicating the formation of adsorption layers near the pore wall. The density of C_1 and nC_4 molecules at the pore center in the 1-nm pore is significantly higher than the density at the center in the 3-nm pore. In smaller pores, the adsorbed molecules at the pore center is much tighter due to the enhanced association forces from both

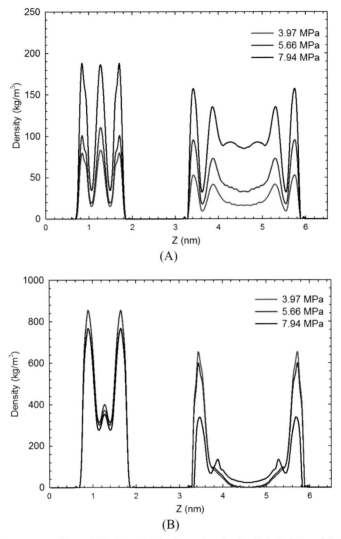

Fig. 3.3 Density profiles of (A) C_1 and (B) nC_4 molecules in C_1/nC_4 (70 mol%:30 mol%) mixture in 1- and 3-nm pores at 333.15 K and three different pressures.

sides of the pore, resulting in a much higher density at the central location of the pore (Liu and Wilcox, 2012; Chen et al., 2016; Liu et al., 2018a,b). Compared to C_1, nC_4 molecules tend to more readily saturate the nanopores due to the stronger surface attractions, resulting in higher adsorption densities in the adsorption layers. It indicates that nC_4 molecules have a higher adsorption capacity on organic pore surface than C_1 molecules. This observation agrees well with the previous experimental measurements (Liu et al., 2018a,b).

In the 3-nm pore, C_1 molecules form one stronger adsorption layer and the second relatively weaker layers. The adsorption density of C_1 molecules in the second adsorption layer is slightly lower than that in the first adsorption layer. However, adsorption density of the second adsorption layer formed by nC_4 molecules is much lower than that of the first stronger adsorption layer. C_1 molecule has a smaller molecular diameter than nC_4, which results in that the adsorption layer formed by C_1 molecules stay much closer to pore surface than that formed by nC_4 molecules. In the 1–nm pore, in addition to the surface adsorption, large proportion of C_1 molecules are highly packed at the pore center, while fewer nC_4 molecules will stay at the pore center; this behavior is probably caused by the competitive adsorption between C_1 and nC_4 molecules on organic pore surface.

As the system pressure drops, the adsorption of C_1 decreases in both pores, but the nC_4 adsorption is becoming stronger in both pores. It is because as pressure decreases, a significant amount of C_1 molecules are released and more spare space is created, rendering nC_4 molecules being with stronger affinity to the pore surface at lower pressures. It thus causes the enhanced adsorption of nC_4 in organic pores as pressure decreases. In previous studies, similar trend has also been found for other adsorbate materials, such as zeolites and pillared layered materials (Lu et al., 2003; Li et al., 2007). Their results imply that as reservoir pressure decreases during shale gas production, lighter components, i.e., C_1, are more readily produced, while the heavier components, i.e., nC_4, tend to stay within nanopores and become difficult to be recovered.

(2) **Competitive adsorption behavior of C_1/CO_2 mixture**. In Fig. 3.4, the density distributions of an equimolar mixture of C_1/CO_2 are presented in the two nanopores at 333.15 K and three different pressures. Compared to CO_2, C_1 molecules exhibit a more pronounced second adsorption layer under the same pressure. It was found that the second adsorption layer is caused by the clustering of molecules (Jin and Firoozabadi, 2013). Therefore, the clustering of C_1 molecules should be much stronger than that of CO_2 molecules, which enhances the adsorption of C_1 molecules in the second adsorption layer.

As pressure drops, density of C_1 in the adsorption layers as well as at the pore center decreases in both nanopores. A similar trend is also observed for CO_2 molecules. Compared to C_1, the density decrement

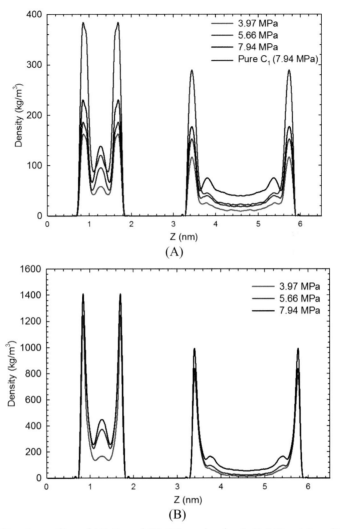

Fig. 3.4 Density profiles of (A) C_1 and (B) CO_2 molecules in C_1/nC_4 mixture (50 mol%: 50 mol%) in 1- and 3-nm pores at 333.15 K and three different pressures.

of CO_2 due to pressure drop is less pronounced. Adsorption capacity of species highly correlates with the system pressure: Adsorption capacity is generally lowered as pressure decreases. When pressure is lowered, the previously adsorbed C_1 and CO_2 molecules are released from pore surface. With the release of the adsorbed molecules, more adsorption sites are liberated, which, on the other hand, enhances the adsorption capacity of other species. Therefore, the competitive adsorption between

C_1 and CO_2 is mainly affected by the coupling effects of pressure and the liberated adsorption sites due to desorption. This coupling effect leads to a relatively less pronounced decrease in the density of the CO_2 adsorption layer.

At 7.94 MPa, pure C_1 molecules are placed in the double-nanopore system and obtain the density profiles in the two pores at 333.15 K. The in situ density of pure C_1 in the double-nanopores is significantly higher than that of C_1 in C_1/CO_2 mixture. CO_2 has stronger adsorption capacity than C_1 on the organic pore surface; thereby, when CO_2 is introduced into the organic pores, C_1 can be replaced by CO_2, especially in the 1-nm pore. Based on this finding, it can be inferred that CO_2 injection is helpful for C_1 recovery from organic pores.

(3) **Competitive adsorption behavior of nC_4/CO_2 mixture**. Fig. 3.5 presents the density profiles of nC_4 and CO_2 in an equimolar mixture of nC_4/CO_2 mixture in the double nanopores at 333.15 K and three different pressures. nC_4 and CO_2 molecules generally show multilayer adsorption in the 3-nm pore; this observation is in line with the previous study conducted in single pores (Liu et al., 2018a,b). However, in the 1-nm pore, nC_4 and CO_2 molecules will exhibit single-layer adsorption; meanwhile, the density at the pore center is high due to the packing effect. Moreover, the adsorption density of nC_4 in the equimolar nC_4/CO_2 mixture is significantly higher than CO_2, indicating its stronger adsorption capacity than CO_2. Fig. 3.5 also presents the density profile of pure nC_4 in the double-nanopores at 7.94 MPa. Density of the adsorption layer of pure nC_4 is higher than that of nC_4 in the nC_4/CO_2 mixture in both nanopores, especially in the 1-nm pore. That is, with the presence of CO_2, adsorption of pure nC_4 is reduced due to the competitive adsorption between nC_4 and CO_2 molecules on pore surface.

As pressure decreases, density of the adsorption layer of CO_2 in the nC_4/CO_2 mixture decreases. Interestingly, density of the adsorption layer of nC_4 in nC_4/CO_2 mixture increases as pressure drops. When pressure decreases, CO_2 molecules desorb from the organic pores surface, liberating more adsorption sites on the pore surface. Although a decrease in the system pressure weakens the adsorption of nC_4, adsorption of nC_4 is enhanced as more liberated adsorption sites are created due to CO_2 desorption. It thereby results in an increasing adsorption of nC_4 molecules in nanopores as pressure decreases. Thereby, when CO_2 is introduced, it may not be effective to use the depressurization approach to recover nC_4 from organic pores.

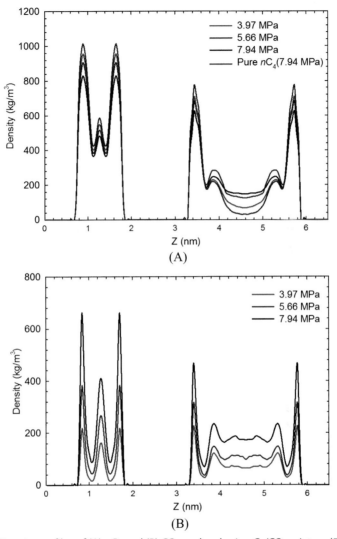

Fig. 3.5 Density profiles of (A) nC_4 and (B) CO_2 molecules in nC_4/CO_2 mixture (50 mol%: 50 mol%) in 1- and 3-nm pores at 333.15 K and three different pressures.

In Fig. 3.6, we show the snapshots of molecular distributions of CH_4/CO_2 (50 mol%:50 mol%) and nC_4/CO_2 (50 mol%:50 mol%) mixtures in the double-pore system at 333.15 K and at 5.66 MPa. As shown in Fig. 3.6A, when CO_2 molecules are introduced into the double-nanopore system which is initially saturated with C_1, CO_2 molecules strongly adsorb on the pore surface due to the interactions between CO_2 and pore surface. On the contrary, C_1 molecules are replaced by

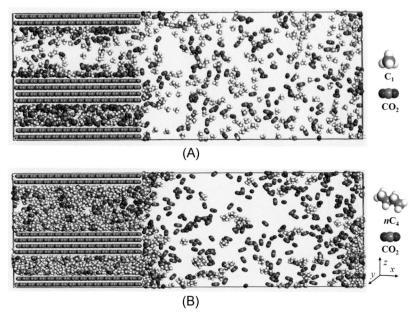

Fig. 3.6 Snapshots of the molecule distributions of (A) CH_4/CO_2 (50 mol%:50 mol%) and (B) nC_4/CO_2 (50 mol%:50 mol%) mixtures in the double-nanopore system at 333.15 K and 5.66 MPa.

CO_2 molecules and mainly appear in the pore center and the bulk space. It is found that the residing molecules are packed more tightly in the 1-nm pore than that in the 3-nm pore due to the confining effect. In Fig. 3.6B, CO_2 molecules are introduced into the double-nanopore system which is initially saturated with nC_4 molecules. Compared to CH_4/CO_2 mixture, CO_2 molecules in the nC_4/CO_2 mixture mainly appear in the pore center and the bulk space. Based on the above results, we can infer that the introduction of CO_2 to C_1-rich shale reservoirs can result in a relatively higher concentration of C_1 in the bulk space than in the nanopores. In comparison, the introduction of CO_2 to nC_4-rich shale reservoirs can result in a relatively higher concentration of CO_2 in the bulk space than in the nanopores. The higher concentration of the lighter component in the bulk space tends to yield a higher bubble point or dew point for the fluids in the bulk space.

(4) **Competitive adsorption behavior of $C_1/nC_4/CO_2$ mixture**. In this subsection, the competitive adsorption behavior of $C_1/nC_4/CO_2$ mixture is further investigated in the double-nanopore system based on the consideration that fluid in shale reservoirs generally comprises of multiple hydrocarbons. Fig. 3.7 presents the density profiles of

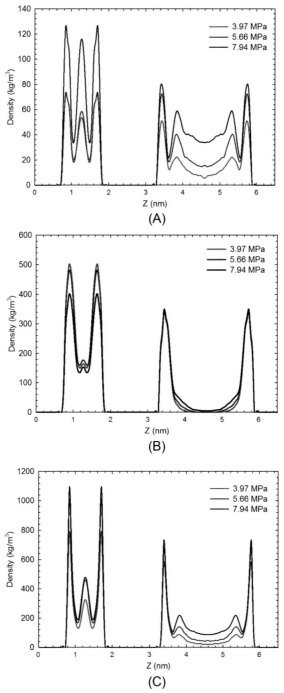

Fig. 3.7 Density profiles of (A) C_1, (B) nC_4, and (C) CO_2 molecules in $C_1/nC_4/CO_2$ (35 mol%: 15 mol%:50 mol%) mixture in 1- and 3-nm pores at 333.15 K and three different pressures.

C_1, nC_4, and CO_2 in $C_1/nC_4/CO_2$ mixture (35 mol%:15 mol%:50 mol%) in the two nanopores at 333.15 K and three different pressures. As shown in Fig. 3.7, CO_2 exhibits the strongest adsorption capacity in both pores, followed by nC_4 and C_1. The higher adsorption capacity of CO_2 than nC_4 is mainly caused by the dominance of CO_2 molecules in the $C_1/nC_4/CO_2$ mixture. As observed from Fig. 3.7B, nC_4 in the $C_1/nC_4/CO_2$ mixture mainly adsorb on the pore surface, exhibiting single-layer adsorption. However, C_1 and CO_2 molecules can form a second weaker adsorption layer in addition to the first stronger adsorption layer. The different adsorption behavior is resulted from the competitive adsorption capacity among C_1, nC_4, and CO_2 molecules on the organic pore surface. As pressure decreases, a significant reduction in the adsorption density is observed for C_1 and CO_2 molecules. Interestingly, as the system pressure decreases, the adsorption density of nC_4 in the 3-nm pore is hardly changed, while the adsorption density of nC_4 in the 1-nm pore is significantly enhanced. During the depressurization, C_1 and CO_2 molecules can be suddenly released from nanopores, while nC_4 molecules would not be produced from the organic pores. It further validates the former findings that CO_2 can efficiently recover C_1 from organic pores but tends to be less efficient for the nC_4 recovery.

3.1.3 Adsorption selectivity of species in organic pores

Figs. 3.8–3.10 show the adsorption selectivity of nC_4 over C_1 (in the C_1/nC_4 mixture), C_1 over CO_2 (in the C_1/CO_2 mixture), and nC_4 over CO_2 (in the CO_2/nC_4 mixture) in the 1- and 3-nm pores, respectively. As shown in Fig. 3.8, adsorption selectivity of nC_4 over C_1 is always higher than 1, suggesting that adsorption capacity of nC_4 on organic pore surface is stronger than that of C_1. This is due to the fact that nC_4 molecules have stronger affinity to the organic surface than C_1 as pressure decreases. In both pores, the adsorption selectivity of both C_1 and nC_4 increases as pressure decreases. Moreover, adsorption selectivity for either C_1 or nC_4 in the 1-nm pore is always higher than that in the 3-nm pore.

Fig. 3.9 shows that the adsorption selectivity of C_1 over CO_2 is always lower than 1, indicating that CO_2 molecule has stronger adsorption capacity on pore surface than C_1 molecule. In the 3-nm pore, adsorption selectivity first increases and then decreases with a decreasing pressure. This observation is in line with the findings by Kurniawan et al. (2006). However, in the

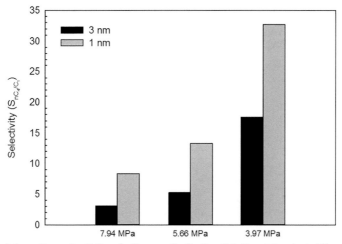

Fig. 3.8 Adsorption selectivity of nC_4 over C_1 (in the C_1/nC_4 mixture) at different pressures in 1- and 3-nm pores.

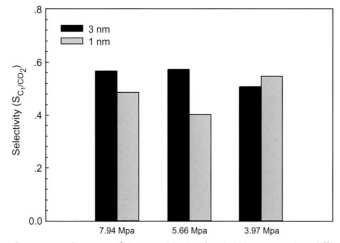

Fig. 3.9 Adsorption selectivity of C_1 over CO_2 (in the C_1/CO_2 mixture) at different pressures in 1- and 3-nm pores.

1-nm pore, adsorption selectivity initially decreases to a minimum value and then increases with a decreasing pressure. As shown in Fig. 3.10, the adsorption selectivity of nC_4 over CO_2 is always higher than 1; it suggests that adsorption capacity of nC_4 is stronger than CO_2. Again, the adsorption selectivity in the 1-nm pore is observed to be higher than that in the 3-nm pore.

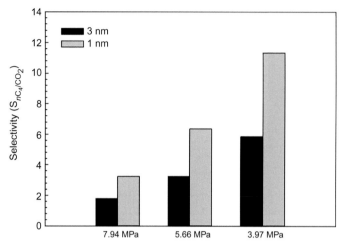

Fig. 3.10 Adsorption selectivity of nC_4 over CO_2 (in the CO_2/nC_4 mixture) at different pressure in 1- and 3-nm pores.

3.1.4 Replacement of C_1 and nC_4 from nanopores with CO_2 injection

In order to see how CO_2 effectively replaces C_1 and nC_4 from organic nanopores, the molar fractions of C_1 and nC_4 in the C_1/CO_2 and nC_4/CO_2 mixtures in the 1- and 3-nm pores are calculated at different pressures. Fig. 3.11 presents the molar fractions of C_1 in C_1/CO_2 mixture in both nanopores at three different pressures. As pressure decreases, molar fraction of C_1 decreases in both nanopores, implying that, as pressure decreases, CO_2 expels more C_1 molecules from organic pores. Moreover, the molar fraction of C_1 in the 1–nm pore is lower than that in the 3–nm pore, suggesting CO_2 is more effective in recovering C_1 molecules from smaller pores. Fig. 3.12 presents molar fractions of nC_4 in nC_4/CO_2 mixture in both nanopores at three different pressures. We can observe that the molar fraction of nC_4 in the two pores is not affected by the system pressure, which is quite different that for C_1. Thereby, in shale industry, CO_2 is a suitable agent to recover lighter hydrocarbons (i.e., C_1) from organic pores, but would not be efficient in recovering heavier hydrocarbons (i.e., nC_4) because the recovery efficiency is strongly affected by the competitive adsorption behavior between hydrocarbons and CO_2 on the organic pore surface.

This study is expected to inspire new understanding on the competitive adsorption of shale hydrocarbons in organic pores and new interpretation of the mechanisms of shale hydrocarbon recovery using CO_2 method.

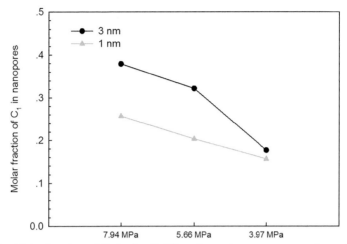

Fig. 3.11 Molar fraction of C_1 in the C_1/CO_2 mixture in nanopores at three different pressures.

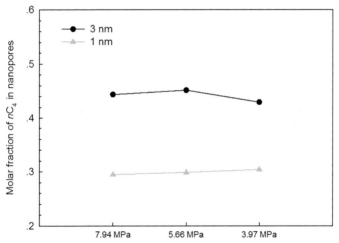

Fig. 3.12 Molar fraction of nC_4 in the nC_4/CO_2 mixture in the nanopores at three different pressures.

Generally, shale fluids may also contain some other hydrocarbons as well as nonhydrocarbons besides methane and butane, such as ethane, propane, and H_2O, etc. In the future works, it is suggested to investigate the adsorption behavior of fluids possessing a full shale-gas composition in nanopores. In addition, shale reservoirs have some unique characteristics, such as extremely-low permeability and heterogeneity; future works should consider how to build nanopore systems with heterogeneity and to reveal the effect of heterogeneity on fluid phase behavior in confined space. In this study, we select 1-nm pore, representing micropore, and 3-nm pore, representing mesopore, to represent the nanopore systems in shale matrix. However, shale samples usually possess pores exhibiting pore-size distribution; specifically, pore size of the pores will continuously vary in shale samples. Thereby, in the future work, it is necessary to explore the adsorption behavior of shale fluids in porous media with full pore-size distribution.

3.2 Determination of the absolute adsorption/ desorption isotherms of CH_4 and n-C_4H_{10} on shale from a nanoscale perspective

Shale resources (such as shale gas or shale gas condensate) have emerged as a key energy resource in recent years. Shale rocks generally have higher total organic carbon (TOC) content than the conventional ones, resulting in hydrocarbons being more apt to adsorb on shale surface (Ross and Bustin, 2009). Thereof, a significant proportion of reserves in shale reservoirs can be in the adsorbed state. During the production of shale gas or shale gas condensate, desorption plays an important role. Adsorption/ desorption of hydrocarbons usually exhibits an interesting phenomenon of hysteresis, and the knowledge about the adsorption/desorption behavior of hydrocarbons in shale is crucial for estimating the hydrocarbon storage capacity and understanding the mechanisms of the subsequent hydrocarbon recovery.

Adsorbed hydrocarbons can account for 20–85 vol% of the total reserves in shale reservoirs (Wu et al., 2015a,b). Many previous researches focused on investigating the adsorption capacity of hydrocarbons on shale rocks (Gasparik et al., 2012; Duan et al., 2016; Wang et al., 2016a,b). CH_4, known to be the most abundant component in shale gas reservoirs, was mostly studied. Some heavier hydrocarbons, e.g., C_2H_6, C_3H_8, and n-C_4H_{10}, can be also present with a large quantity in shale reservoirs, up to 20 vol% (Wang et al., 2015). But adsorptions of these heavier components in shale rocks are scarcely measured in the literature. Pedram et al. (1984) measured the adsorption isotherms of C_2H_6, C_3H_8, and n-C_4H_{10} in two oil–shale samples and found that n-C_4H_{10} has the highest adsorption capacity, followed by C_3H_8 and C_2H_6. But it is noted that the oil-shale they used still have residual oil left in the samples, which can affect the gas adsorption on shale due to the large solubility of various hydrocarbons in shale oil. Therefore, such measured adsorption isotherms could not represent the actual adsorption capacity of gases on shale. Recently, Wang et al. (2015) measured the excess adsorption isotherms of pure CH_4 and C_2H_6 on shale samples. C_2H_6 is shown to have a higher adsorption capacity than CH_4, and Wang et al. (2015) attributed this finding to that C_2H_6 is more apt to get adsorbed on shale samples than CH_4. But this conclusion is made based on the measured excess adsorption isotherms, rather than the absolute adsorption isotherms; excess adsorption isotherms are generally not accurate enough as it neglects the adsorbed-phase volume occupied by the adsorbed gas.

By knowing the pore volume from the helium adsorption, volumetric method is commonly used to measure the adsorption isotherms of hydrocarbons on shale samples (Heller and Zoback, 2014; Gasparik et al., 2014). Recently, some scholars used the thermogravimetric analysis (TGA) technique to measure the adsorption isotherms (Wang et al., 2015). Compared with the volumetric method, TGA loads a smaller sample amount into the setup; the magnetic suspension balance mounted in the TGA setup is capable of measuring the weight change down to 1 µg, rendering the TGA technique more accurate than the volumetric method. However, the adsorption isotherms directly measured by TGA technique are excess adsorption isotherms, which neglects the adsorbed-phase volume and thereby underestimates the total adsorption amount. The density of the adsorption phase is commonly used to correct the excess adsorption isotherms, yielding the absolute adsorption isotherms. In the adsorption phase, gas molecules are in an adsorbed state; to our knowledge, few efforts are dedicated to quantifying the density of the adsorption phase. Previously, constant density values are normally used to pragmatically represent the density of the adsorption phase. Dubinin (1960) suggested that the density of the adsorption phase is a constant value which correlates with the van der Waals constant *b*. Later, the density of adsorption phase is argued to be equal to the liquid adsorbate density (Wang et al., 2016a,b; Menon, 1968; Tsai et al., 1985). Li et al. (2002) compared the aforementioned methods and claimed that the density of the adsorption phase is a function of the system temperature, but its value approaches that proposed by Dubinin (1960). Recently, with molecular simulations, Ambrose et al. (2012) suggested that the density of the adsorption phase correlates with the system temperature, pressure, and pore size. Actually, fluids in confined space are strongly affected by fluid/pore-surface interactions, especially in shale samples which are usually abundant in nanoscale pores. It is, thereby, of critical importance to precisely capture the density of the adsorption phase in order to more accurately determine the absolute adsorption isotherms.

In this section, the objectives are multifold: (1) to use GCMC simulations to capture the in situ density distribution in carbon-slit pores under the effects of the system pressure, temperature, and pore size; (2) to determine the absolute adsorption/desorption isotherms of hydrocarbons on shale samples by knowing the in situ density of the adsorption phase; and (3) to further analyze and compare the characteristics of the absolute adsorption/desorption isotherms of CH_4 and n-C_4H_{10}. As part of a comprehensive study on the adsorption/desorption behavior of hydrocarbons in shale reservoirs, we measure the adsorption/desorption isotherms of CH_4 and n-C_4H_{10} on two shale samples using the TGA technique, and then determine the absolute adsorption/desorption isotherms based on GCMC simulations. CH_4 is selected with the consideration that CH_4 is the most abundant component in shale gas, while n-C_4H_{10} adsorption/desorption isotherms are measured to represent the adsorption/desorption behavior of heavier hydrocarbons in shale reservoirs.

3.2.1 Characterization of shale sample

This section presents the procedures used to characterize the shale samples as well as the characterization results. Various techniques, including the TOC measurement, the scanning election microscopy (SEM), and the N_2 adsorption/desorption test are adopted to characterize the shale samples.

The TOC content of two shale samples is measured by a combustion elemental analyzer. In this measurement, the organic carbon in shale samples are sparged with oxygen, forming carbon dioxide; then the TOC content is determined by detecting the amount of the carbon dioxide with the non-dispersive infrared detector. The TOC contents of the two shale samples are shown in Table 3.2. We observe shale sample #1 has a higher TOC content of 3.71 wt%, 3.78 times of that in shale sample #2. The measured TOC contents are in good agreement with the reported values for Longmaxi shale which range from 0.52 to 6.05 wt% (Wu et al., 2015a,b).

The Hitachi TM-300 SEM setup is used to characterize the surface morphology at an accelerating voltage of 20.0 kV. Prior to scanning, shale surface is polished with argon ion. Subsequently, the polished shale surface is coated

Table 3.2 TOC contents and BET surface areas of the two shale samples used in this study.

Shale sample ID	TOC content (wt%)	R_o (%)	BET surface area (m²/g)
#1	3.71	2.35	2.98
#2	0.98	1.82	2.06

Fig. 3.13 The FE-SEM images of the two shale samples. Energy-dispersive X-ray spectroscopy (EDX) analysis has been conducted at the sites marked by *"a"* and *"b."*

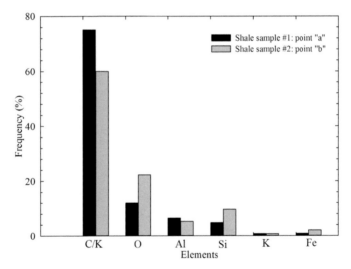

Fig. 3.14 Energy-dispersive X-ray spectroscopy (EDX) analysis results for points *"a"* and *"b"* shown in Fig. 3.13.

with a golden film with a thickness of 10 nm to improve the conductivity. Fig. 3.13 shows the FE-SEM images taken on the two shale samples. The energy–dispersive X-ray spectroscopy (EDX) analysis is further conducted on the chosen points "a" and "b" in shale samples #1 and #2, respectively, as marked in Fig. 3.13. Fig. 3.14 shows the EDX test results. As seen from Fig. 3.14, a high concentration of carbon element is present at both sites, indicating that organic matter, i.e., kerogen, is residing in both sites.

It can be also seen from Fig. 3.13 that the kerogen is surrounded by meso-pores, which is a typical characteristic observed for kerogen in shale.

Pore-size distribution and Brunauer-Emmett-Teller (BET) surface area are characterized by the N_2 adsorption/desorption tests conducted with the Autosorb iQ-Chemiadsorption and Physi-adsorption Gas Adsorption Analyzer (Quantachrome Instruments, United States). In a wide range of testing pressure, N_2 adsorption isotherms can characterize pore-size distributions in the micro-, meso-, and macroporosity range (approximately 0.5–200 nm) (Martin and Siepmann, 1998). Therefore, in view of the nature of our shale samples, we select N_2 as the adsorptive to obtain the PSD of two shale samples. Fig. 3.15 presents the pore-size distribution of the two shale samples as obtained by analyzing the N_2 isotherm data measured at 77.0 K with the nonlocal density functional theory (NLDFT). With the NLDFT method, the whole region of micro- and mesopores can be characterized, although there are still some issues that cannot be addressed by this method (Errington and Panagiotopoulos, 1999). For instance, the NLDFT method cannot account for the networking effects and transition from the models of independent pores to the pore networks (Errington and Panagiotopoulos, 1999). The dominant pore size of shale sample #1 is around 4.2 nm, while the dominant pore size of shale sample #2 is around 3.3 nm. Shale sample #1 possesses more mesopores (2–50 nm) and macropores (larger than 50 nm) than shale sample #2, indicating a higher thermal maturity of the organic matter in shale sample #1. The R_o value is further measured for each shale sample, while such R_o value represents the thermal maturity of organic matter in shale samples. The R_o values for shale samples #1 and #2 are 2.35% and 1.82%, respectively, which validate the our former statement. As measured in this work, the BET surface area obtained for shale sample #1 is higher than that for shale sample #2.

3.2.2 Excess and absolute adsorption/desorption

The excess adsorption/desorption isotherms of CH_4 and n-C_4H_{10} are measured using a thermogravimetric analyzer (TGA) (IEA-100B, Hiden Isochema Ltd., United Kingdom). The key component of TGA is a magnetic suspension balance with 1.0 μg accuracy in weight measurement. The test pressures are set up to 50 bar for CH_4 and up to 2 bar for n-C_4H_{10}, respectively, while the test temperatures are set at 303.15, 333.15, 368.15, and 393.15 K. An electrical heater is applied to keep a constant temperature during the adsorption/desorption measurements.

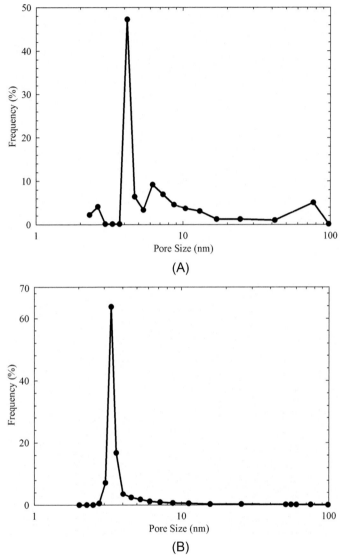

Fig. 3.15 Pore-size distributions of (A) shale sample #1 and (B) shale sample #2 as obtained from N$_2$ adsorption/desorption test.

It should be noted that 2 bar is the highest pressure value we can reach due to the low vapor pressure of n-C$_4$H$_{10}$ at room temperature. Each test is repeated twice to make sure the measured results are reliable and reproducible. The maximum deviation between two consecutive runs is found to be less than ±1.56%.

With TGA technique, the measured excess adsorption uptake (M_{ex}) is obtained by (Wang et al., 2015),

$$M_{ex} = M_a - \rho V_a = M_{app} - (M_s + M_{sc}) + \rho(V_s + V_{sc}) \tag{3.5}$$

where M_a is the adsorbed uptake on the shale sample, which is defined as the absolute adsorption uptake (M_{abs}), kg; ρ is the bulk gas density, kg/m^3; V_a is the adsorption-phase volume, m^3; M_{app} is the apparent weight measured by TGA, kg; M_s and M_{sc} are the weight of shale sample and the weight of the sample container, respectively, kg; and ($V_s + V_{sc}$) is the total volume of the shale sample and the sample container, m^3.

It has been found that, when pore size is large enough, the gas density in the pore center approaches that in bulk (Ambrose et al., 2012). Thereof, in nanopores, the distribution of CH_4 or $n\text{-}C_4H_{10}$ molecules can be divided into free-gas region and adsorption-phase region. Fig. 3.16 schematically shows the absolute adsorption uptake, the excess adsorption uptake, the free-gas region, and the adsorption-phase region in a nanopore. As shown in Fig. 3.16, the density of the adsorption phase (ρ_{ads}) is higher than bulk

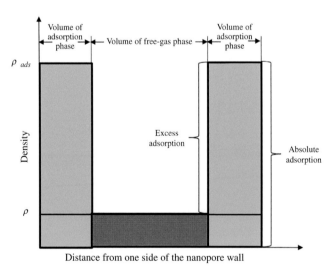

Fig. 3.16 Schematic of the absolute adsorption and excess adsorption in nanopores. ρ_{ads} is the density of the adsorption phase, and ρ is the density of the free-gas phase, which is equal to the bulk gas density.

free-gas phase density (ρ). The green area depicted in Fig. 3.16 shows the absolute adsorption. Based on knowledge of the density of the adsorption phase and absolute adsorption uptake (M_{abs}), the adsorption-phase volume (V_a) can be calculated using the following equation,

$$V_a = \frac{M_{abs}}{\rho_{ads}} \qquad (3.6)$$

Therefore, the actual adsorbed amount on the shale sample, i.e., absolute adsorption uptake, can be obtained by,

$$M_{abs} = \frac{M_{ex}}{1 - \dfrac{\rho}{\rho_{ads}}} \qquad (3.7)$$

Thereof, the key to obtain an accurate absolute adsorption uptake is to accurately calculate the density of the adsorption phase. It is known that the density of the adsorption phase is a function of system pressure, temperature, and pore size (Ambrose et al., 2012). However, in previous works, the density of the adsorption phase was provided as a constant which was either calculated from van der Waals constant b (Dubinin, 1960) or obtained from the liquid density (Wang et al., 2016a,b; Menon, 1968; Tsai et al., 1985). From a nanopore-scale perspective, molecular simulations can faithfully capture the properties of the adsorption phase over a wide pressure and temperature range due to the consideration of fluid/pore-surface interactions. The density of adsorption phase is calculated using the GCMC simulations.

3.2.3 Grand canonical Monte Carlo (GCMC) simulations

Within grand canonical (GC) ensemble, the entire system has fixed volume (V) and temperature (T) and chemical potential (μ). Since the number of molecules in the system fluctuates during the simulations, the average number of molecules in the ensemble is fully determined by the chemical potential.

In this model, the united atom model (Martin and Siepmann, 1998) is used to simulate different hydrocarbon molecules. The modified Buckingham exponential-6 intermolecular potential (Errington and Panagiotopoulos, 1999) is applied to describe nonbonded site-site interactions among functional groups on different molecules, as well as among functional groups belonging to the same molecule separated by more than three bonds. The pairwise interaction potential $U(r)$ for the nonbonded site-site interactions is given as (Errington and Panagiotopoulos, 1999):

$$U(r) = \begin{cases} \dfrac{\varepsilon}{1-\dfrac{6}{\alpha}}\left[\dfrac{6}{\alpha}\exp\left(\alpha\left[1-\dfrac{r}{r_m}\right]\right) - \left(\dfrac{r_m}{r}\right)^6\right], & r > r_{max} \\ \infty, & r < r_{max} \end{cases}$$
(3.8)

where r is the interparticle separation distance, r_m is the radial distance at which $U(r)$ reaches a minimum, and the cutoff distance r_{max} is the smallest radial distance for which $dU(r)/dr=0$ (Singh et al., 2009). Since the original Buckingham exponential-6 potential can be negative at very short distances, the cutoff distance is thus defined to avoid negative potentials (Errington and Panagiotopoulos, 1999). The radial distance at which $U(r)=0$ is defined as σ. The values of the exponential-6 parameters ε, σ, and α are 129.63 K, 0.3679 nm, and 16, respectively, for the methyl group ($-CH_3$), 73.5 K, 0.4 nm, and 22, respectively, for the methylene group ($-CH_2-$), and 160.3 K, 0.373 nm, 15, respectively, for CH_4 (Singh et al., 2009). The cross parameters are determined by the following combining rules (Singh et al., 2009),

$$\sigma_{ij} = \frac{1}{2}\left(\sigma_i + \sigma_j\right)$$
(3.9)

$$\varepsilon_{ij} = \left(\varepsilon_i\varepsilon_j\right)^{1/2}$$
(3.10)

$$\alpha_{ij} = \left(\alpha_i\alpha_j\right)^{1/2}$$
(3.11)

The bond lengths for CH_3-CH_2 and CH_2-CH_2 are taken as 0.1687 nm and 0.1535 nm, respectively. The torsion potential ($U_{tor}(\varphi)$) is expressed as (Smit et al., 1995),

$$U_{tor}(\varphi) = V_0 + \frac{V_1}{2}(1+\cos\varphi) + \frac{V_2}{2}(1-\cos 2\varphi) + \frac{V_3}{2}(1+\cos 3\varphi)$$
(3.12)

where φ is the torsional angle from equilibrium, V_0, V_1, V_2, and V_3 are 0, 355.03, -68.19, and 791.32 K, respectively. The bond bending potential $U_{bend}(\theta)$ is calculated by (van der Ploeg and Berendsen, 1982),

$$U_{bend}(\theta) = \frac{K_\theta}{2}\left(\theta - \theta_{eq}\right)^2$$
(3.13)

where parameter K_θ is equal to 62,500 K/rad^2, θ is the bond angle from equilibrium, and θ_{eq} is the equilibrium bond angle (114 degrees).

It has been found that the higher organic carbon content enables hydro-carbons to be more apt to adsorb on shale surface (Kim et al., 2017a,b). Thereby, in this model, nanopores are selected as slit geometry with smooth and structureless carbon surfaces. 10-4-3 Steele potentials are used to describe the fluid-pore surface interactions φ_{wf} (Steele, 1973),

$$\varphi_{wf}(z) = 2\pi\rho_w\varepsilon_{wf}\sigma_{wf}^2\Delta\left[\frac{2}{5}\left(\frac{\sigma_{wf}}{z}\right)^{10} - \left(\frac{\sigma_{wf}}{z}\right)^4 - \frac{\sigma_{wf}^4}{3\Delta(0.61\Delta + z)^3}\right] \quad (3.14)$$

where z is the distance of the fluid particle from the pore surface, ρ_{wf} is the density of carbon atom per unit surface area of the graphite layer ($114\,nm^{-3}$), The molecular parameters of an atom in the graphite layer are $\varepsilon_{wf} = 28\,K$, and $\sigma_{wf} = 0.3345\,nm$ (Do and Do, 2003), and Δ is the spacing between two adjacent graphene layers (0.335 nm), respectively. The external potential ψ in a slit pore is given as (Steele, 1973),

$$\psi(z) = \varphi_{wf}(z) + \varphi_{wf}(W - z) \quad (3.15)$$

where W is the size of the slit pore.

In each MC cycle, a trial random displacement is applied to all CH_4 molecules; with equal probability, a CH_4 molecule is randomly removed from or inserted into the simulations box depending on the chemical potential of CH_4. For simulations of $n\text{-}C_4H_{10}$ molecules in slit pores, in addition to the MC moves as mentioned above, a trial random rotation is applied to all $n\text{-}C_4H_{10}$ molecules. A configurational-biased GCMC algorithm is used to insert and remove $n\text{-}C_4H_{10}$ molecules (Hensen et al., 2001). The Widom insertion method (Widom, 1963) is used to obtain the chemical potentials of bulk CH_4 and $n\text{-}C_4H_{10}$ molecules in canonical ensemble. The PR-EOS (Peng and Robinson, 1976) is applied to calculate the bulk densities of CH_4 and $n\text{-}C_4H_{10}$ at given pressure and temperature. The MC moves are implemented by using the Metropolis algorithm (Metropolis et al., 1953). During the simulations, 0.1 million of MC cycles per each adsorbate molecule is required to reach an equilibrium state, while 0.5 million of MC cycles per adsorbate molecule is required to sample the density profiles.

The average density (ρ_{ave}) of component i in carbon-slit pores is expressed as,

$$\rho_{ave} = \frac{\langle N_i\rangle M_i}{VN_A} \quad (3.16)$$

where $\langle N_i \rangle$ is the ensemble averaged number of molecules of component i in nanopores, V is the volume, M is molecular weight of components i, and N_A is Avogadro constant, 6.022×10^{23}.

3.2.4 Density distributions in nanopores

To calculate the density of the adsorption phase, density distributions in nanopores should be known a priori. We investigate the density distributions of pure CH_4 or n-C_4H_{10} in a single carbon-slit pore. The effects of the system pressure, temperature, and pore size are examined. In the GCMC framework, CH_4 molecules are regarded as spherical particles, while n-C_4H_{10} molecules are represented considering the orientation and configuration (Jin and Firoozabadi, 2016a,b).

3.2.4.1 Effect of system pressure

With molecular simulations, Ambrose et al. (2012) found that the CH_4 adsorption behavior in nanopores is sensitive to changes in pressure. To illustrate the effect of system pressure on adsorption behavior of CH_4 and n-C_4H_{10}, in Fig. 3.17, the density distributions of CH_4 and n-C_4H_{10} in 4.2 nm pore are presented at 368.15 K and different system pressures. It is noted that the 4.2 nm is the dominant pore size of shale sample #1. At all bulk pressure conditions, both CH_4 and n-C_4H_{10} molecules can form one strong adsorption layer and the density in the pore center approaches the bulk density obtained from NIST (Lemmon et al., 2009). Thereby, the gas in the adsorption layer can be stated as the adsorbed gas, while the gas located in the pore center can be taken as the free gas. As for CH_4, when pressure is larger than 35 bar, a second weak adsorption layer can form in the location adjacent to the first adsorption layer, while n-C_4H_{10} forms such a second adsorption layer when system pressure is larger than 0.4 bar. As the bulk pressure increases, the second adsorption layer becomes more pronounced due to the stronger interactions between molecules, as depicted in Fig. 3.17. Compared with CH_4, the second adsorption layer of n-C_4H_{10} is stronger due to the stronger molecule/molecule interactions. On the contrary, at a relatively lower pressure, CH_4 or n-C_4H_{10} molecules form only one adsorption layer; beyond this adsorption layer, the density is slightly higher than the bulk density, which corresponds to a transition zone in the density profiles (Didar and Akkutlu, 2013; Tian et al., 2017). Furthermore, we observe that, the density of the adsorption layers of CH_4 and n-C_4H_{10} increases with pressure. Therefore, it may not be appropriate to use a constant density value to

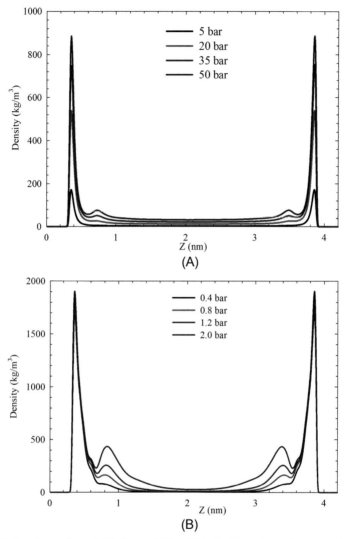

Fig. 3.17 Density profiles of (A) CH_4 and (B) n-C_4H_{10} in the carbon-slit pore of 4.2 nm at 368.15 K and different pressures.

represent the density of the adsorption phase (Wang et al., 2016a,b; Heller and Zoback, 2014).

3.2.4.2 Effect of system temperature

Fig. 3.18 shows the density distributions of CH_4 or n-C_4H_{10} molecules in a carbon-slit pore of 4.2 nm under different system temperatures. As the

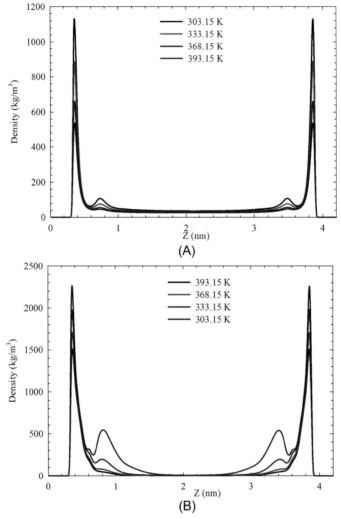

Fig. 3.18 Density profiles of (A) CH_4 in the carbon-slit pore of 4.2 nm at 50 bar and (B) n-C_4H_{10} in the carbon-slit pore of 4.2 nm at 0.4 bar.

system temperature decreases, the density of the adsorption layer increases. However, as temperature increases, adsorption of CH_4 or n-C_4H_{10} is significantly suppressed, which is manifested by the drops in the density of the two adsorption layers; this observation is in line with a previous study by Ambrose et al. (2012). It is due to the weaker fluid/surface interaction at higher temperatures. Comparatively, the density of the two adsorption layers of n-C_4H_{10} is higher than that of CH_4. It is probably because the surface attraction of the carbon wall to n-C_4H_{10} is stronger than that to CH_4, which greatly enhances the adsorption of the heavier alkane, n-C_4H_{10}.

3.2.4.3 Effect of pore size

To reveal the effect of pore size on density profiles, in Fig. 3.19, the density distributions of CH_4 and n-C_4H_{10} molecules are presented in carbon-slit

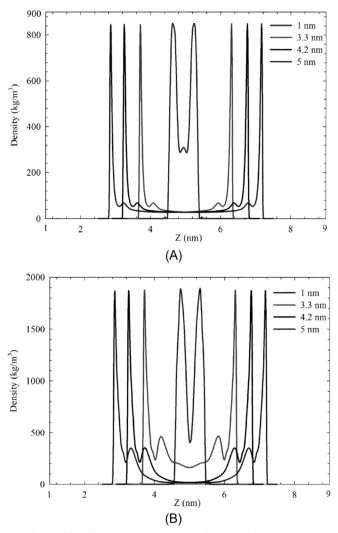

(A)

(B)

Fig. 3.19 Density profiles of (A) CH_4 in the carbon-slit pores of 1.0, 3.3, 4.2, and 5.0 nm at 333.15 K and 45 bar and (B) n-C_4H_{10} in the carbon-slit pores of 1.0, 3.3, 4.2, and 5.0 nm at 368.15 K and 1.6 bar.

pores of 1.0, 3.3, 4.2, and 5.0 nm. In pores with a size larger than 1.0 nm, CH_4 and n-C_4H_{10} molecules can form two adsorption layers, while, in the 1-nm pore, only one adsorption layer forms on the pore surface due to the limited pore space. In addition, the density in the center of 4.2- and 5.0-nm pores approaches the bulk, while the density in the center of 1.0-nm pore is much higher than the bulk value. As the pore size becomes as narrow as 1.0 nm, the packing of molecules in the pore center becomes tighter due to the enhanced attraction forces from the both sides of the pore, leading to the much higher density in the central location of the pore (Liu and Wilcox, 2012; Chen et al., 2016). It indicates that there is no free-gas region in such nanopores. It is interesting to observe from Fig. 3.19 that the density profiles exhibited by CH_4 molecules in the 3.3-nm pore well resemble those in the 4.2- and 5.0-nm pores. It implies that once the pore size is larger than a certain value, a change in the pore size will not affect the configuration of the adsorption layers formed by the CH_4 molecules. As for n-C_4H_{10}, the density of the free-gas phase in 3.3-nm pore is much higher than those in 4.2- and 5.0-nm pores, while the adsorption phase in the 3.3-nm pore well resembles that in the 4.2- and 5.0-nm pores. It is clear that the fluid distributions in nanopores can be greatly affected by the pore size. Our results indicate that fluid distributions of CH_4 and n-C_4H_{10} vary in response to the changes in system pressure, temperature, and pore size.

3.2.4.4 Identification of the adsorption phase

One issue needs to be addressed herein, i.e., how to determine the cutoff distance that separates the free-gas phase and the adsorption phase. As can be observed from Fig. 3.19, in mesopores (2–50 nm), two adsorption layers are formed, and the density in the pore center approaches that in the bulk. However, in micropores (<2 nm), only one adsorption layer is formed and the density in the pore center is much higher than bulk. This observation is in line with the previous study by Tian et al. (2017). As a result, in micropores, it is not justifiable to use the adsorption model in Fig. 3.16. Considering the two studied shale cores mainly contain mesopores, we thus can define the free-gas phase and the adsorption phase.

CH$_4$ adsorption in 4.2-nm carbon-slit pore is used as an example to illustrate the methodology for determining the adsorption phase. Fig. 3.20 presents the density distributions of CH_4 confined in the carbon-slit pore of 4.2 nm at 333.15 K and 50 bar. As shown in this figure, the adsorption phase is defined as the region between a (or a') and b (or b'). The volume between points a and a' is depicted as the all accessible pore volume of the bulk free gas

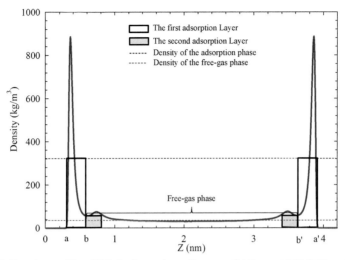

Fig. 3.20 Density profile of CH_4 in the carbon-slit pore of 4.2 nm at 368.15 K and 50 bar.

(Tian et al., 2017). Point b (or b') is the saddle point between the first adsorption layer and the second weak adsorption layer. The width of the adsorption phase of CH_4 (ab), around 0.37 nm, is similar to the diameter of CH_4 molecule. For n-C_4H_{10}, the width of the adsorption phase, around 0.42 nm, is also similar to the diameter of n-C_4H_{10} molecules. It indicates that CH_4 and n-C_4H_{10} generally exhibits single-layered Langmuir adsorption on pore surface under the experimental conditions, which agrees well with the previous studies (Li et al., 2014; Dong et al., 2016). Using this methodology, we can determine the width of the adsorption phase for CH_4 or n-C_4H_{10} in 3.3- and 4.2-nm pores under the experimental pressure/temperature conditions. It is noted that 4.2 and 3.3 nm are the dominant pore sizes of shale sample #1 and #2, respectively. As shown in Fig. 3.17, it is found that, at a given temperature, the width of the adsorption phase remains almost unchanged as the system pressure increases. However, Fig. 3.18 shows that, at a given pressure, the width of the adsorption phase increases as the system temperature increases. At a higher temperature, the larger width is probably resulted from the weaker carbon surface/gas interactions.

3.2.5 Average density of the adsorption phase

By knowing the width of the adsorption phase, the average density of the adsorption phase for CH_4 or n-C_4H_{10} can be thereby calculated by

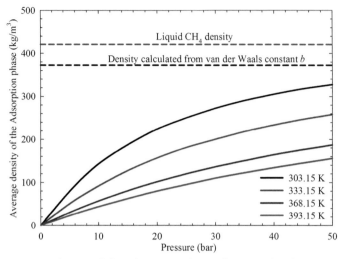

Fig. 3.21 Average density of the adsorption phase of CH_4 confined in the carbon-slit pore of 3.3 and 4.2 nm at different temperatures and pressures. The constant density of liquid CH_4 and the density calculated from van der Waals constant b are also shown in this figure. It should be noted that the average density of the adsorption phase of CH_4 confined in 3.3-nm pore is identical to that in the 4.2-nm pore.

$\rho_{ave} = \int_a^b \rho_{ads}(z) dz / z_{ab}$ (ρ_{ave} is the average density of the adsorption phase; ρ_{ads} is the in situ density of the adsorption phase; and z_{ab} is the distance between a and b) (see Fig. 3.20). Figs. 3.21 and 3.22 show the average density of the adsorption phase for CH_4 and $n\text{-}C_4H_{10}$, respectively; the density of the adsorption phase of CH_4 or $n\text{-}C_4H_{10}$ are calculated in carbon–slit pores of 3.3 and 4.2 nm at different pressures and temperatures. The average density of the adsorption phase of CH_4 or $n\text{-}C_4H_{10}$ in 3.3-nm pore is identical to that in 4.2-nm pore. Furthermore, as for CH_4 and $n\text{-}C_4H_{10}$, the average density of the adsorption phase strongly correlates with the system pressure and temperature: it increases as the system pressure increases (or as the system temperature decreases).

Fig. 3.21 also shows the density of liquid CH_4, 421 kg/m^3 (Wang et al., 2016a,b), and another constant density of CH_4 calculated from the van der Waals constant b (Dubinin, 1960; Li et al., 2002; Rexer et al., 2013). It is noted that the liquid CH_4 density has been extensively used as the density of the adsorption phase to obtain the absolute adsorption isotherms (Wang et al., 2016a,b) or fit empirical models to the adsorption isotherms (Rexer et al., 2013; Weniger et al., 2010). The constant value of 421 kg/m^3 is mostly

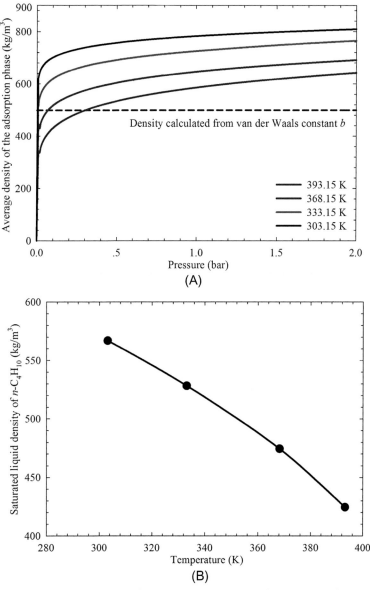

Fig. 3.22 (A) Average density of the adsorption phase of n-C_4H_{10} confined in the carbon-slit pore of 3.3, and 4.2 nm at different temperatures and pressures: The constant density of liquid n-C_4H_{10} calculated from van der Waals constant b is also shown in this figure; (B) Saturated liquid density of n-C_4H_{10} as a function of temperature is calculated by Eq. (3.13). It should be noted that the average density of the adsorption phase of n-C_4H_{10} confined in 3.3-nm pore is identical to that in the 4.2-nm pore.

used. The constant density of CH_4 based on the van der Waals constant b is also heavily used to represent the density of the adsorption phase, i.e., $1/b$ – (Dubinin 1960; Li et al., 2002; Rexer et al., 2013). Fig. 3.22A also shows the density of liquid n-C_4H_{10} calculated from the van der Waals constant b, $502 \, kg/m^3$. Since the saturated liquid density of n-C_4H_{10} is known to correlate with system temperature, the following correlation can be used to calculate the saturated liquid density of n-C_4H_{10}, as depicted in Fig. 3.22B (Yaws, 2003),

$$\log_{10}\left(\frac{\rho_{nC_4H_{10}}}{1000}\right) = \log_{10}(h) - \log_{10}(l)\left(1 - \frac{T}{T_c}\right)^n \qquad (3.17)$$

where ρ_{nC_4H10} is saturated density of n-C_4H_{10}, kg/m^3; T_c is the critical temperature of n-C_4H_{10}; h, l, and n are coefficients with values of 0.2283, 0.2724, and 0.2863, respectively. The critical pressure and temperature of n-C_4H_{10} used are listed in Table 3.3. It is clear that, as for either CH_4 or n-C_4H_{10}, the density of the adsorption phase should be a variable which depends on the in situ temperature/pressure, rather than a constant value.

As shown in Fig. 3.20, the free gas is defined as the region between points b and b', which covers the second weak adsorption layer. In Fig. 3.23, we compare the average density of the free-gas region of CH_4 in a 4.2-nm pore calculated by the GCMC simulations with the bulk density from NIST (Lemmon et al., 2009). The average density of the free-gas region for CH_4 is calculated by $\rho_f = \int_b^{b'}\rho(z)dz/z_{bb'}$ (ρ_f is the average density of the free-gas phase; ρ is the in situ density of the free-gas phase; and $z_{bb'}$ is the distance between b and b') (see Fig. 3.20). Fig. 3.23 shows the comparative results at 333.15 K, demonstrating that the density values calculated from GCMC simulations is in a good agreement with the NIST data, especially at relatively low pressures. This proves the reliability of the GCMC simulations. But deviation shows up at pressures larger than 30 bar and increases as pressure further increases. Such deviation can be attributed to the presence of the transition zone (Tian et al., 2017) (See Fig. 3.17).

Table 3.3 Critical properties of n-C_4H_{10} used for density calculation (Firoozabadi, 2016).

Adsorbate	T_c (K)	P_c (bar)
n-C_4H_{10}	425.18	37.97

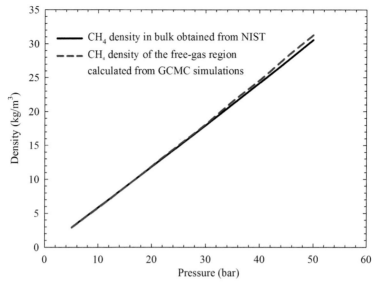

Fig. 3.23 Comparisons of CH_4 density of the free-gas region at the carbon-slit pore of 4.2 nm at 333.15 K calculated by GCMC simulations with CH_4 density in bulk obtained from NIST.

3.2.6 Absolute adsorption/desorption isotherms

Since the measured adsorption/desorption isotherms are surface excess quantities, the density of the adsorption phase is required to transform these excess values to absolute ones. Based on the density of the adsorption phase computed from GCMC simulations, the excess adsorption/desorption isotherms are converted to absolute ones. Figs. 3.24 and 3.25 present the converted absolute adsorption/desorption isotherms of CH_4 and n-C_4H_{10} on the two shale samples studied. The absolute adsorption of CH_4 or n-C_4H_{10} increases as pressure increases or as temperature decreases. At the same pressure and temperature, n-C_4H_{10} has relatively higher adsorption capacity compared to CH_4; it is because pore surface shows stronger attractions toward n-C_4H_{10} molecules than CH_4, indicating a higher affinity of n-C_4H_{10} toward shale. In shale reservoirs, the heavier hydrocarbons can be more easily to get adsorbed on the shale surface, forming liquid-phase-like structures and showing stronger storage capacity as the adsorbed state (Li et al., 2014).

The difference in the adsorption and desorption isotherms is termed as the hysteresis phenomenon. This hysteresis behavior can be attributed to the capillary condensation taking place in nanopores as pressure changes at a

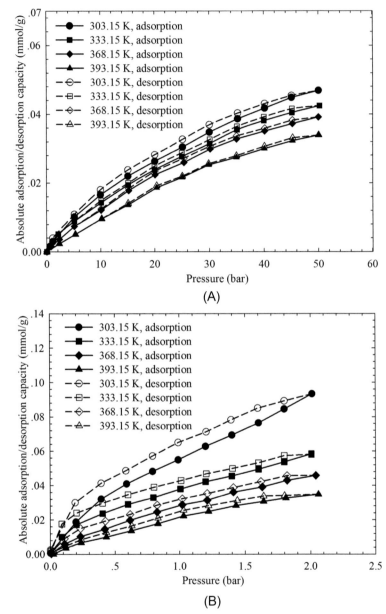

Fig. 3.24 Absolute adsorption/desorption isotherms of (A) CH_4 and (B) n-C_4H_{10} on shale sample #1. These isotherms are obtained by converting the excess adsorption/desorption isotherms based on the average density of the adsorption phase calculated by GCMC simulations.

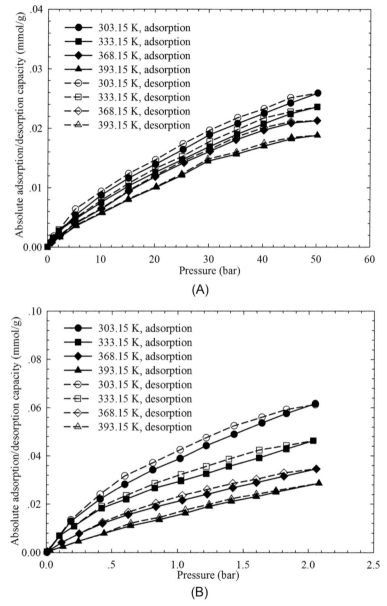

Fig. 3.25 Absolute adsorption/desorption isotherms of (A) CH_4 and (b) n-C_4H_{10} on shale sample #2. These isotherms are obtained by converting the excess adsorption/desorption isotherms based on the average density of the adsorption phase calculated by GCMC simulations.

given temperature (Li et al., 2014; Bryan, 1987). With density functional theory (DFT), Li et al. (2014) studied the adsorption/desorption hysteresis of pure CH_4 and pure n-C_4H_{10} in a single carbon-slit pore and found that the hysteresis phenomenon for pure component only occurs over a small pressure range at a given temperature. The measured results shown in Figs. 3.25 and 3.26 show that, however, in a real shale sample, the hysteresis phenomenon for CH_4 or n-C_4H_{10} appears over the entire pressure range at a given temperature. The real shale samples are porous medium, in which pore-size distribution is presented. The hysteresis in shale samples is not as sharp as that in a carbon-slit pore because shale samples comprise of different pore sizes that may show hysteresis at different pressures. As for both CH_4 and n-C_4H_{10}, the hysteresis phenomenon is getting more pronounced at a lower temperature. Comparatively speaking, n-C_4H_{10} exhibits stronger adsorption/desorption hysteresis than CH_4, which agrees well with the simulation studies based on the use of DFT (Li et al., 2014).

Comparing Figs. 3.24 with 3.25, CH_4 or n-C_4H_{10} exhibits a higher adsorption capacity on shale sample #1 than shale sample #2. Adsorption strongly correlates with the TOC content and surface area in the shale sample (Kim et al., 2017a,b). Thereof, such higher adsorption on shale sample #1 may be caused by the higher TOC content (3.17 wt%) and larger BET surface area (2.98 m^2/g) than those of shale sample #2 (a TOC content of 0.98 wt% and a BET surface area of 2.06 m^2/g). However, Xiong et al. (2017) presented that the adsorption capacity does not correlate only with the TOC and surface area but shows a more complex dependence on the petrophysical and mineralogical properties; therefore, to understand the adsorption capacity of CH_4 or n-C_4H_{10}, more adsorption data should be measured and other factors, e.g., clay minerals, should be considered to understand the adsorption behavior.

In Fig. 3.26, the excess adsorption isotherms of CH_4 and n-C_4H_{10} are compared against the corresponding absolute adsorption isotherms. As mentioned above, the excess adsorption isotherms are converted to the absolute adsorption isotherms using the in situ density of the adsorption phase which are calculated from GCMC simulations. It can be seen from Fig. 3.26A that, as for CH_4, the absolute adsorption is found to be always higher than the directly measured excess adsorption. A relatively large deviation is found to exist between the absolute adsorption isotherms and the excess adsorption isotherms for CH_4, which highlights the importance of using accurate density of the adsorption phase to obtain accurate absolute adsorption isotherms for CH_4. However, as for n-C_4H_{10}, the absolute adsorption isotherms are

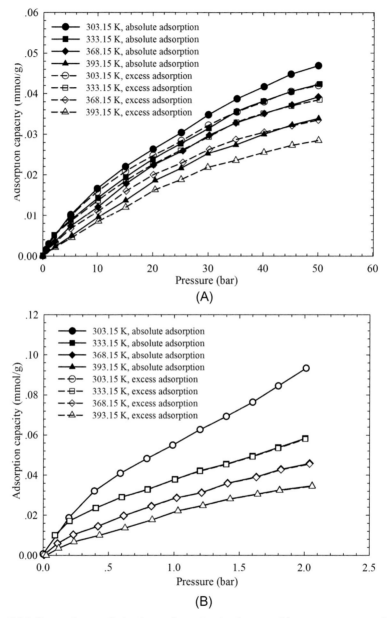

Fig. 3.26 Comparisons of absolute adsorption isotherms with excess ones on shale sample #1: (A) CH_4 and (B) n-C_4H_{10}. The absolute adsorption isotherms have been converted from the excess adsorption isotherms based on density of the adsorption phase which is calculated by GCMC simulations.

almost identical to the excess adsorption isotherms (see Fig. 3.26B). In this work, the adsorption of n-C$_4$H$_{10}$ is measured on shale samples at pressures only up to 2 bar; under such low pressures, the bulk gas density is far less than the density of the adsorption phase, as seen from Fig. 3.17B. As a result, the term ρ/ρ_a is a value approaching zero. As such, the denominator in the right hand side of Eq. (3.7) approaches 1, rendering the absolute adsorption being almost equal to the excess adsorption. This explains why the absolute adsorption isotherms for n-C$_4$H$_{10}$ are almost identical to the excess adsorption isotherms, as shown in Fig. 3.26B.

3.2.7 Comparison of GCMC-based approach with conventional approach

The liquid density and the density calculated from van der Waals constant b are commonly used to approximate the density of the adsorption phase. Herein, the densities calculated from these two conventional approaches are used to convert the measured excess adsorption isotherms to the absolute ones. Thereafter, the absolute adsorption isotherms converted by the two conventional approaches are compared with those calculated from the GCMC simulations. Fig. 3.27 compares the absolute adsorption capacity of CH$_4$ and n-C$_4$H$_{10}$ on shale sample #1 calculated by GCMC-based approach against that calculated by using the liquid density of CH$_4$ or n-C$_4$H$_{10}$, while Fig. 3.28 compares the absolute adsorption capacity of CH$_4$ and n-C$_4$H$_{10}$ on shale sample #1 calculated by GCMC-based approach against that calculated using the van der Waals constant b. As can be seen from Figs. 3.27B and 3.28B, as for n-C$_4$H$_{10}$, the conventional approaches and the GCMC-based approach provide almost the same conversion results. However, as seen from Figs. 3.27A and 3.28A, the two conventional approaches tend to underestimate the absolute adsorption for CH$_4$. These aforementioned findings highlight the importance of obtaining an accurate estimation of the adsorption-phase density, especially when one wants to accurately evaluate the total amount of gas–in–place in shale gas reservoirs.

Although this work may provide an alternative to correct the measured excess adsorption, there are still some issues needed to be addressed in the future work. Firstly, the adsorption isotherms are only measured at the pressures up to 2 bar for nC$_4$H$_{10}$ considering the low vapor pressure of nC$_4$H$_{10}$ at room temperature. It is the drawback in this measurement. If possible, the testing pressures should be as high as the reservoir conditions, which would be more interesting to engineers. Secondly, the adsorption/desorption isotherms are measured for CH$_4$ and n-C$_4$H$_{10}$. However, various hydrocarbon

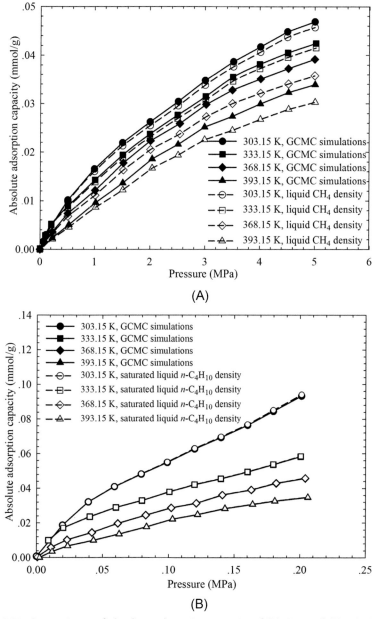

Fig. 3.27 Comparisons of absolute adsorption capacity of (A) CH_4 and (B) n-C_4H_{10} on shale sample #1 calculated by GCMC-based approach with that obtained by the liquid density of CH_4 or n-C_4H_{10}.

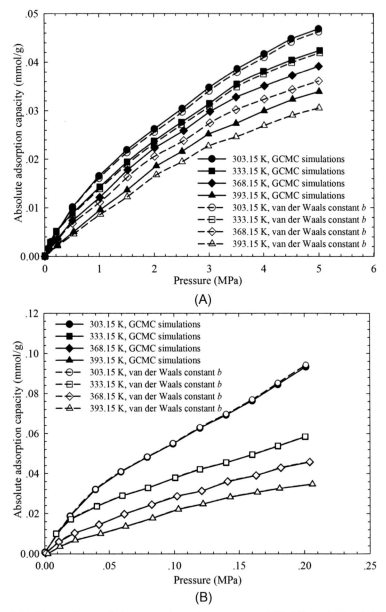

Fig. 3.28 Comparisons of absolute adsorption capacity of (A) CH_4 and (B) n-C_4H_{10} on shale sample #1 calculated by GCMC-based approach with that obtained by the density calculated by van der Waals constant b.

components (C_2H_6, and C_3H_8) and nonhydrocarbon components (CO_2 and N_2) may coexist in shale gas; thereby, more adsorption measurements should be conducted on the other components except CH_4 and $n\text{-}C_4H_{10}$ to have a comprehensive understanding of the adsorption behavior of components in shale reservoirs. Additionally, shale gas is multicomponent system, in which different components present distinct adsorption behavior on shale, exhibiting selective adsorption. Therefore, to better estimate shale gas-in-place storage, excess adsorption isotherms of gas mixtures should be measured and new techniques should be developed to correct the excess ones and obtain the accurate absolute adsorption of gas mixtures on shale.

3.3 Absolute adsorption of CH_4 on shale with the simplified local density theory

Shale gas, as one kind of unconventional energy, is playing an important role in supplying energy resources in recent years. Comparing to conventional reservoirs, shale reservoir has unique characteristics (Liu et al., 2018a,b); such as shale rocks contains large amount of organic matter. Organic matter mainly comprises of kerogen, which provides a significant amount of nanopores. Due to the large proportion of kerogen as well as nanopores, shale gas can adsorb on organic pore surface, which is critical in accessing the gas storage in shale reservoirs (Ross and Bustin, 2009). Understanding of shale-gas adsorption on organic shale is crucial in assessing the storage of shale resources and thus revealing the mechanisms of shale fluid transportation in shale matrix.

CH_4 in natural gas produced from shale is one of the most commonly seen components in shale gas (Dasani et al., 2017); generally, it exists in shale reservoirs in three states, namely, the adsorbed state by the organic walls, the free-gas state in pores, and the absorbed-gas state in kerogen (Clarkson and Haghshenas, 2013; Wu et al., 2015a,b; Huang et al., 2019). It has been found that significant amount of CH_4 may be stored in the adsorbed state (Cristancho-Albarracin et al., 2017). Volumetric method and TGA method are extensively employed to measure the excess CH_4 adsorption on shale samples (Gasparik et al., 2012; Heller and Zoback, 2014; Wang et al., 2016a,b). Volumetric method measures the excess adsorption by deducting the total amount of gas in free state in all available pore volume from the measured total gas uptake (Lu et al., 1995; Tian et al., 2017). TGA method obtains the excess adsorption uptake by measuring the force difference between gravity and buoyancy using a magnetic suspension balance

(Zhou et al., 2016). Comparing to the volumetric method, TGA can measure the weight difference down to 1 µg, which is more accurate in measuring the adsorption uptake than the volumetric method.

Laboratory adsorption measurement can only provide the excess adsorption data, which is taken as the difference between the total amount of CH_4 in organic pores and the amount of free CH_4 in all accessible pore volume (Fitzgerald, 1999). Actually, adsorption of CH_4 correlates with the petrophysical properties of shale cores, which may be strongly affected by the surface properties of organic pore walls (Cristancho-Albarracin et al., 2017). Santos and Akkutlu (2013) designed a gas-storage measuring method to combine porosity with sorption measurements considering that the routine measurements are generally performed separately. Their method successfully takes into consideration of the influence of adsorption layer and pore compressibility on the measured sorption data.

To evaluate the adsorbed amount of CH_4, the measured excess adsorption is generally converted to the absolute value, which reflects the total adsorbed CH_4 on shale (Tian et al., 2017; Liu et al., 2018a,b). In previous works (Rexer et al., 2013), the conversion between the excess adsorption uptake (M_{ex}) and the absolute adsorption uptake (M_a) is expressed as,

$$M_a = \frac{M_{ex}}{1 - \dfrac{\rho}{\rho_a}} \tag{3.18}$$

where ρ_a represents the adsorbed CH_4 density; ρ represents the bulk CH_4 density. We obtain the bulk CH_4 density experimentally. Thereby, an accurate determination of the adsorbed CH_4 density is critical to obtain the absolute adsorption.

In previous studies, the adsorbed CH_4 density is generally taken as constant value. Dubinin (1960) proposed that the adsorbed CH_4 density is related with the van der Waals parameter b, which is calculated as a constant value (Dubinin, 1960). Recently, the density of adsorbed CH_4 is represented by liquid CH_4 density at the normal boiling point, i.e., 420 kg/m³ (Wang et al., 2016a,b). To date, it was proved that the adsorbed CH_4 density varies with pressure and temperature (Liu et al., 2018a,b; Jin and Firoozabadi, 2016a,b; Ambrose et al., 2012; Riewchotisakul and Akkutlu, 2016). The modified Langmuir adsorption model, Dubinin-Radushkevich (DR) equation, Ono-Kondo, and Supercritical DR (SDR) models can be used to obtain the absolute CH_4 adsorption by matching the measured excess data (Gensterblum et al., 2009; Ambrose et al., 2012; Xiong et al., 2017). However, these models match the excess adsorption data by adjusting the adsorbed CH_4 density, which are only curve-fitting methods without any physical mechanisms.

CH_4 adsorption is also investigated using the molecular simulation methods. Compared to these curve-fitting methods, molecular simulations provide the underlying mechanisms of CH_4 adsorption in nanopores. Using Monte Carlo simulations, Jin and Firoozabadi (2016a,b) observed that surface area dominates CH_4 adsorption in nanopores. In our previous work, we determined the absolute CH_4 adsorption using the grand canonical Monte Carlo (GCMC) simulations (Liu et al., 2018a,b). By considering the CH_4-surface interactions, the adsorbed CH_4 density calculated from GCMC simulations is found to change with pressure and temperature, which is in agreement with the previous observation by Ambrose et al. (2012). Although molecular simulation methods determine the absolute CH_4 adsorption with higher accuracy than the conventional methods, the computation is sometimes expensive. Compared with molecular simulations, SLD theory simplifies the fluid-pore wall interactions as well as the chemical potential calculations, rending it more efficient than the molecular simulations in obtaining the density distribution of fluids in pores. Moreover, by specifically considering the fluid/surface interactions, density distribution of fluid in carbon-slit pores calculated from SLD theory is heterogeneous, which agrees reasonably well with the results from GCMC simulations. Thereby, SLD theory could be served as an alternative to the molecular simulations in calculating the adsorbed density of CH_4 in organic pores.

In this section, the excess CH_4 adsorption is measured on four shale samples. Based on the measured excess adsorption, SLD theory is employed to calculate the corresponding absolute CH_4 adsorption. Meanwhile, SLD model is validated by comparing with the GCMC simulations. The main objective is to propose an efficient approach to accurately determine the absolute CH_4 adsorption on shale. Based on the SEM imaging, it is observed that pore structures in shale are mainly slit-shaped, while it also includes cylindrical and ink-bottle shapes (de Boer and Lippens, 1964; Sing et al., 2008). Based on the GCMC simulation method, previous studies have proposed that pore size as well as the pore shape can result in a difference in the adsorption isotherms due to the difference in the adsorbate-adsorbate interaction energy, and the adsorbate-pore interaction energy (Keffer et al., 1996; Dobrzanski et al., 2018). Future work should be conducted to explore the effect of pore-size distribution and pore geometry on the adsorption behavior of hydrocarbons in nanopores. In the SLD model, a carbon-slit pore model is used to describe the nanopores in shale samples for simplicity. In this regard, Ambrose et al. (2012) also used carbon–slit pore to represent kerogen walls in shale, considering that carbon surface is hydrocarbon-wet and can provide underlying mechanisms on the CH_4 adsorption in nanopores.

3.3.1 Characterization of the four shale samples

In this subsection, the procedures are presented for characterizing the four shale samples. N_2 adsorption/desorption BET test is adopted to obtain the pore-size distribution (PSD), and porosity of the four shale samples; scanning election microscopy (SEM) test is conducted to characterize the surface morphology of the four shale samples. We also present the characterization results in this section.

The N_2 adsorption/desorption tests are used to characterize the PSD, porosity as well as the Brunauer-Emmett-Teller (BET) surface area of shale samples. These tests are carried out with the Gas Adsorption Analyzer (Quantachrome, United States). The N_2 adsorption data measured at 77.0 K are analyzed with the nonlocal density functional theory (NLDFT) to obtain the PSD (Tarazona et al., 1987), while the specific surface area is calculated using the BET equation (Gregg and Sing, 1982). The NLDFT model is enabled to characterize the whole region of micro- and mesopores, i.e., <2 and 2–50 nm (Groen et al., 2003). However, the NLDFT model suffers the following drawbacks, such as the swelling effects, the networking effects, and the transition from the models of independent pores to the pore networks cannot be considered (Groen et al., 2003; Liu et al., 2018a,b). Fig. 3.29 shows the measured N_2 adsorption/desorption isotherms of the four shale samples. Fig. 3.30 presents the measured PSD of the four shale samples, while Table 3.4 shows the measured BET surface area of the four shale samples. It is found that pores in these samples are generally in nanoscale with pore size in the range of 1 to 100 nm. As shown in Fig. 3.30, the dominant pore sizes of four shale samples are 2.23, 3.25, 4.20, and 6.21 nm, respectively.

The surface morphology of the four shale samples are characterize with the Hitachi SEM setup. Before the SEM scanning, shale surface is firstly polished with the argon ion. To improve the conductivity, the ion–polished shale surface is consequently covered with a golden film, of which the thickness is about 10.0 nm. Fig. 3.31 shows the digital SEM images obtained for the four shale samples. Nanoscale kerogen pores are presented in these shale samples. By conducting the energy dispersive X-ray spectroscopy (EDX)

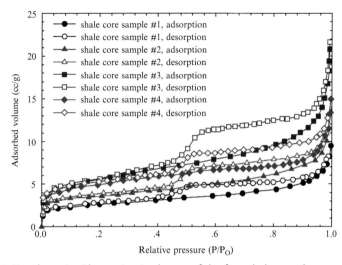

Fig. 3.29 N_2 adsorption/desorption isotherms of the four shale samples.

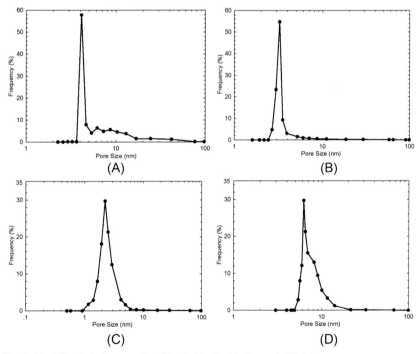

Fig. 3.30 PSD of shale samples (A) #1, (B) #2, (C) #3, and (D) #4.

Table 3.4 The measured porosity, TOC content, and BET surface area of the four shale samples.

Sample ID	Porosity (%)	BET surface area (m^2/g)
#1	1.80	19.69
#2	3.45	37.03
#3	2.94	25.69
#4	2.03	16.32

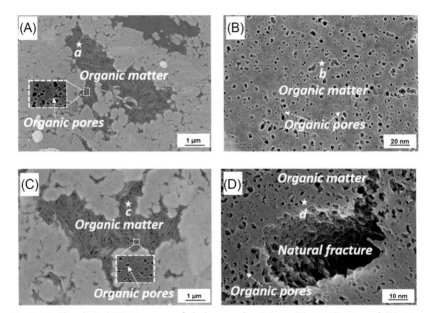

Fig. 3.31 The digital SEM images of shale samples (A) #1, (B) #2, (C) #3, and (D) #4.

analysis on these shale samples (as depicted as *a*, *b*, *c*, and *d* in Fig. 3.31), a high carbon concentration is exhibited at the four sites; it indicates that kerogen is residing on these sites.

3.3.2 Simplified local density (SLD) theory

The SLD theory has been employed to describe the CH$_4$ adsorption on carbon-slit pores (Rangarajan et al., 1995). The equation of state of CH$_4$ applied in the SLD model is simplified with a local–density approximation in obtaining the configurational energy of the adsorbed CH$_4$ which is in-homogeneously distributed in nanopores (Mohammad et al., 2009). This model explicitly takes into consideration the CH$_4$–CH$_4$ and CH$_4$-pore surface interactions, and thus can reasonably describe the adsorption behavior of fluid in nanoscale pores. Comparing to molecular simulations, SLD can significantly reduce the computational cost. Thereby, it could be a supplementary tool to other available engineering models.

To develop the SLD theory, three main assumptions are proposed as follows (Mohammad et al., 2009),

(1) Near the pore surface, the chemical potential at any position equals to the chemical potential in bulk;

(2) In nanopores, the chemical potential at any position is calculated as the summation of CH_4–CH_4 and CH_4-solid surface potentials when adsorption reaches equilibrium;

(3) The fluid-solid surface potentials at any position is not affected by the number of molecules at or around this position.

When adsorption equilibrium is achieved, the chemical potential of CH_4 at position z is given by the summation of the potentials of CH_4–CH_4 and CH_4-solid surface interactions, which equals to the chemical potential of CH_4 in bulk.

$$\mu(z) = \mu_{ff}(z) + \mu_{fs}(z) = \mu_{bulk} \tag{3.19}$$

where the subscript "ff" represents the CH_4–CH_4 interactions, "fs" represents CH_4-solid surface interactions, and "$bulk$" represents CH_4 in bulk.

The chemical potential of CH_4 in bulk is calculated as a function of fugacity, which is expressed as,

$$\mu_{bulk} = \mu_0(T) + RT \ln\left(\frac{f_{bulk}}{f_0}\right) \tag{3.20}$$

where f_{bulk} represents the fugacity of CH_4 in bulk, f_0 represents the fugacity at an arbitrary reference state. In the same way, the chemical potential of CH_4 in carbon-slit pore due to the CH_4–CH_4 interaction is expressed as,

$$\mu_{ff}(z) = \mu_0(T) + RT \ln\left(\frac{f_{ff}(z)}{f_0}\right) \tag{3.21}$$

where $f_{ff}(z)$ represents the fugacity of CH_4 at the position z; and f_0 is the fugacity at the same arbitrary reference state as that in Eq. (3.20).

The chemical potential of CH_4 in a pore due to the CH_4-solid surface interaction is given as (Rangarajan et al., 1995),

$$\mu_{fs}(z) = N_A\left[\Psi^{fs}(z) + \Psi^{fs}(L-z)\right] \tag{3.22}$$

where $\Psi^{fs}(z)$ and $\Psi^{fs}(L-z)$ represent the CH_4-solid surface interactions from both solid surfaces of a carbon-slit pore with a pore size of L; N_A is Avogadro's number. The CH_4-solid surface interaction is given by the Lee's partially integrated 10-4 Lennard-Jones potential (Lee, 1988).

$$\Psi^{fs}(z) = 4\pi\rho_{atoms}\varepsilon_{fs}\sigma_{fs}^2\left(\frac{\sigma_{fs}^{10}}{5(z')^{10}} - \frac{1}{2}\sum_{i=1}^{4}\frac{\sigma_{fs}^4}{(z'+(i-1)\sigma_{ss})^4}\right) \qquad (3.23)$$

where ρ_{atoms} represents the solid-atom density, 38.2 atoms/nm^2 (Mohammad et al., 2011); ε_{fs} represents the parameter due to the CH_4-solid surface interaction; σ_{fs} is the CH_4-solid surface molecular diameter, which is calculated by $\sigma_{fs} = (\sigma_{ff} + \sigma_{ss})/2$, where σ_{ff} and σ_{ss} are the molecular diameter of CH_4 and the carbon-interplanar distance, respectively. As for graphite, the value of σ_{ss} is 0.355 nm; z' is the dummy coordinate, $z' = z + \sigma_{ss}/2$.

The criterion for adsorption equilibrium yields by substituting Eqs. (3.6)–(3.8) into Eq. (3.4) as,

$$f_{ff}(z) = f_{bulk}\exp\left(-\frac{\psi^{fs}(z) + \psi^{fs}(L-z)}{kT}\right) \qquad (3.24)$$

where k represents the Boltzmann's constant, 1.38×10^{-23} J/K; and T represents the absolute temperature.

The Peng-Robinson equation of state (PR-EOS) is applied to calculate CH_4–CH_4 interaction. The PR-EOS can be written as a function of density (ρ) as,

$$\frac{P}{\rho RT} = \frac{1}{(1-\rho b)} - \frac{a(T)\rho}{RT[1+(1-\sqrt{2})\rho b][1+(1+\sqrt{2})\rho b]} \qquad (3.25)$$

where

$$a(T) = \frac{0.457535\alpha(T)R^2T_c^2}{P_c} \qquad (3.26)$$

$$b = \frac{0.077796\alpha(T)RT_c}{P_c} \qquad (3.27)$$

The $\alpha(T)$ term in Eq. (3.26) is given with the following expression (Gasem et al., 2001).

$$\alpha(T) = \exp\left[(A+BT_r)\left(1-T_r^{C+D\omega+E\omega^2}\right)\right] \qquad (3.28)$$

where A, B, C, and D represent correlation parameters, which are 2.0, 0.8145, 0.508, and -0.0467, respectively. As for CH_4, the value of acentric factor (ω), the critical pressure (P_c), the critical temperature (T_c), and the molecular diameter are 0.0113, 4.6 MPa, 190.56 K, and 0.3758 nm, respectively.

In the PR-EOS, we can calculate the fugacity of bulk CH_4 as,

$$\ln\frac{f_{bulk}}{P} = \frac{b\rho}{1-b\rho} - \frac{a(T)\rho}{PT(1+2b\rho-b^2\rho^2)} - \ln\left[\frac{P}{RT\rho} - \frac{Pb}{RT}\right]$$
$$- \frac{a(T)}{2\sqrt{2}RT}\ln\left[\frac{1+\left(1+\sqrt{2}\right)\rho b}{1+\left(1-\sqrt{2}\right)\rho b}\right] \tag{3.29}$$

where P is the pressure of fluid in bulk. With a similar analogy, the fugacity of the adsorbate due to the CH_4–CH_4 interaction is given as,

$$\ln\frac{f_{ff}(z)}{P} = \frac{b\rho(z)}{1-b\rho(z)} - \frac{a_{ads}(z)\rho(z)}{PT(1+2b\rho(z)-b^2\rho^2(z))} - \ln\left[\frac{P}{RT\rho(z)} - \frac{Pb}{RT}\right]$$
$$- \frac{a_{ads}(z)}{2\sqrt{2}bRT}\ln\left[\frac{1+\left(1+\sqrt{2}\right)\rho(z)b}{1+\left(1-\sqrt{2}\right)\rho(z)b}\right] \tag{3.30}$$

where the term $a_{ads}(z)$ depends on the position in the carbon-slit pore and the dimensionless pore width L/σ_{ff} (Chen et al., 1997). The equations for calculating the $a_{ads}(z)$ are proposed by Chen et al. (1997) and are presented in the Appendix 1. $\rho(z)$ represents the in situ gas density in nanopores, which depends on the position in carbon-slit pores (Chen et al., 1997).

In the PR-EOS, the covolume parameter b can significantly affect the local density of the adsorbed CH_4 near the solid surface (Mohammad et al., 2009). Fitzgerald (2005) modified the covolume parameter b to enhance the predictive capacity of pure CH_4 on carbon surface (Fitzgerald, 2005). To account for the repulsive interactions of adsorbed CH_4 at high pressure conditions, the covolume parameter b is modified by the parameter Λ_b as (Fitzgerald, 2005),

$$b_{ads} = b(1+\Lambda_b) \tag{3.31}$$

where b_{ads} represents the modified covolume; Λ_b represents the empirical correction, which generally ranges from -0.4 to 0.0 for shale gases (Mohammad et al., 2009). In our model, this value is fixed at -0.20 for CH_4. After this modification, Eq. (3.30) is written as,

$$\ln \frac{f_{ff}(z)}{P} = \frac{b_{ads}\rho(z)}{1 - b_{ads}\rho(z)} - \frac{a_{ads}(z)\rho(z)}{PT\left(1 + 2b_{ads}\rho(z) - b_{ads}^2\rho^2(z)\right)} - \ln\left[\frac{P}{RT\rho(z)} - \frac{Pb_{ads}}{RT}\right]$$

$$- \frac{a_{ads}(z)}{2\sqrt{2}b_{ads}RT} \ln\left[\frac{1 + \left(1 + \sqrt{2}\right)\rho(z)b_{ads}}{1 + \left(1 - \sqrt{2}\right)\rho(z)b_{ads}}\right]$$

$$(3.32)$$

The density distribution of CH_4 in a nanopore can be obtained by combining Eqs. (3.19) through (3.32). Within the SLD theory, the excess CH_4 adsorption is expressed as,

$$n^{ex} = \frac{A}{2} \int_{\frac{\sigma_{ff}}{2}}^{L - \frac{\sigma_{ff}}{2}} [\rho(z) - \rho_{bulk}]dz \qquad (3.33)$$

where n^{ex} represents the excess CH_4 adsorption in moles per unit mass of adsorbent; A represents the total surface area of adsorbed CH_4 on the carbon surface. The lower limit of integration $\sigma_{ff}/2$ represents the center of the sphere-shaped CH_4 molecule adsorbed on the pore surface, while the upper limit of integration $L - (\sigma_{ff}/2)$ is the center of CH_4 molecule adsorbed on the pore surface of the other wall.

The average density (ρ_{ave}) of CH_4 in nanopores is calculated by,

$$\rho_{ave} = \frac{\int_0^W \rho(z)dz}{W} \qquad (3.34)$$

where W represents the pore size of carbon-slit pore.

In this work, the SLD model is applied to predict the CH_4 distribution, i.e., density distribution, in nanopores at different pressure/temperature conditions. Based on the predicted density distribution, the excess adsorption of CH_4 is thus calculated using the Eq. (3.33), which is used to match the directly measured excess CH_4 adsorption from the TGA method. By adjusting the key parameters in the SLD model, i.e., fluid–pore surface interaction energy, ε_{fs}/k, and covolume correction parameter, b_{ads}, a reasonable agreement can be achieved between the measured excess adsorption and that predicted from the SLD model. The SLD model is employed again to calculate the CH_4 distribution in nanopores at given temperature/pressure conditions; and the adsorbed CH_4 density is then calculated based on the CH_4 distribution, which is applied in the Eq. (3.18) to convert the measured excess adsorption to the absolute adsorption values.

3.3.3 Density distributions of CH_4 in nanopores

Fig. 3.32 presents the density profiles of CH_4 in a 2.23-nm pore at different pressure conditions. Note that the pore size of 2.23 nm has the most probability in shale sample #1. The in situ density near the pore surface is significantly higher than that at the pore center; moreover, only one peak density is observed at all pressure conditions, indicating that CH_4 exhibits single layered adsorption on the organic pore surface. This observation shows a good agreement with the previous studies (Dong et al., 2016; Li et al., 2014). It is observed that the density of CH_4 in the adsorption layer increases as pressure increases, suggesting that CH_4 adsorption is affected by the system pressure. Previously, Riewchotisakul and Akkutlu (2016) investigated the steady-state flow of CH_4 in carbon nanotubes using the nonequilibrium-molecular-dynamic simulation method. Similarly, they observed that the density profile of CH_4 varies with the system pressure; in addition, it was found that the density of CH_4 in the adsorbed phase is also influenced by the system pressure. At the pore center, the CH_4 density approaches the bulk density of CH_4. Interestingly, at lower-pressure conditions, CH_4 density at the pore center is slightly higher than that in bulk, while it is almost identical to the

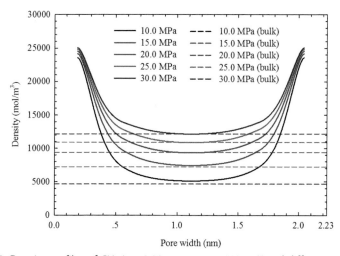

Fig. 3.32 Density profiles of CH_4 in a 2.23-nm pore at 303.15 K and different pressures. The bulk CH_4 density is obtained from NIST Chemistry WebBook (Lemmon et al., 2009).

bulk density at higher pressures. In a nanopore, fluid–fluid and fluid–surface interactions vary with the system pressures, which may result in different density profiles at the pore center.

In Fig. 3.33, the density profiles of CH_4 in the 2.23-nm pore are shown at different temperature conditions. As temperature increases, CH_4 density in the adsorption layer decreases. It suggests that CH_4 adsorption is suppressed at high temperature conditions, which is similar to the previous studies which were conducted using GCMC simulations (Liu et al., 2018a,b; Ambrose et al., 2012). In other words, the density of CH_4 in the adsorption-layer is also influenced by the system temperature. At low temperatures, CH_4 density at the pore center is somewhat higher than that in bulk, while such density is almost similar to the bulk density at high temperatures. In a nanopore, fluid distribution at the pore center is mainly resulted from the fluid–fluid interaction, which is, however, a function of the system temperature; that is, the system temperature can affect the fluid distribution at the pore center by changing the fluid–fluid interaction.

Pore walls can attract fluid molecules residing in nanopores and thus affect fluid distributions in pores. It suggests that pore size may impose impact on fluid distribution in nanopores. Fig. 3.34 shows the density profiles of CH_4 molecules in nanopores with different pore sizes, i.e., 1.0, 2.0, 3.0, 4.0, and 5.0 nm. It is found that the density profiles of CH_4 vary in pores

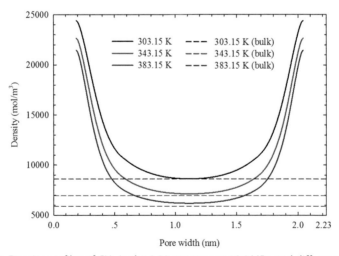

Fig. 3.33 Density profiles of CH_4 in the 2.23-nm pore at 18.0 MPa and different temperatures. The bulk CH_4 density is obtained from NIST Chemistry WebBook (Lemmon et al., 2009).

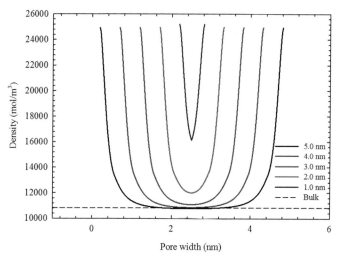

Fig. 3.34 Density profiles of CH_4 molecules in the nanopores with different pore sizes, at 343.15 K and 30.0 MPa. The bulk CH_4 density at 343.15 K and 30.0 MPa is obtained from NIST Chemistry WebBook (Lemmon et al., 2009).

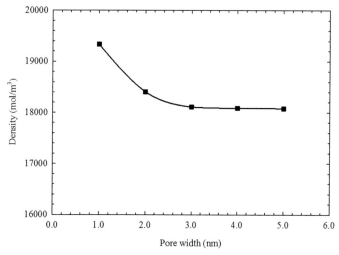

Fig. 3.35 The average CH_4 density in the adsorption layer in pores with different pore sizes.

with different pore sizes. Fig. 3.35 presents the average CH_4 density in the adsorption layer in these nanopores. The average density in the adsorption layer increases as pore size decreases due to the strong attraction from pore surface. Thereby, it is inferred that pore size can affect the adsorption-phase density of CH_4 in nanopores.

The bulk CH$_4$ density is obtained at 343.15 K and 30.0 MPa from NIST Chemistry WebBook and depicted in Fig. 3.34. It is found that CH$_4$ density at the pore center in larger nanopores, i.e., 4.0- and 5.0-nm nanopores, is approaching the density of CH$_4$ in bulk, while it is higher than the bulk CH$_4$ density in smaller nanopores, i.e., 1.0-, 2.0-, and 3.0-nm nanopores (Lemmon et al., 2009). When pore size becomes smaller, CH$_4$ molecules at the pore center tend to tightly compact due to the improved attraction from both pore surfaces (Liu and Wilcox, 2012; Chen et al., 2016). It results in a much higher CH$_4$ density at the pore center, suggesting that no free-gas can be observed in smaller pores. It can be inferred that pore size cannot only affect fluid distribution in the adsorption phase but also affect fluid distribution in the free-gas phase. Overall, the results calculated from SLD theory indicates that fluid distribution of CH$_4$ in nanopores is correlated with temperature, pressure, and pore size.

To calculate the adsorbed CH$_4$ density, the obtained density profile is divided into three regions, i.e., adsorption phase, transition zone, and free-gas phase, as depicted in Fig. 3.36. The adsorption phase of CH$_4$ in nanopores is recognized as the region located between points a (or a') and b (or b'), while the total volume between points a and a' is taken as all available pore volume of CH$_4$ molecules (Tian et al., 2017). The region between points c and c' is defined as the free-gas phase, where the density in this region approaches the bulk density. As shown in Fig. 3.36, the region lying between the adsorption phase and the free-gas phase is depicted as the

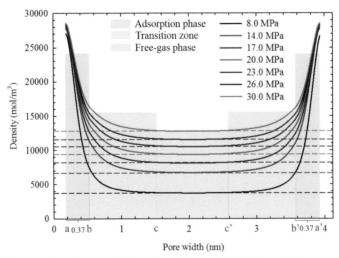

Fig. 3.36 Density distributions of CH$_4$ in the 4.20 nm at 333.15 K and different pressures.

transition zone (Didar and Akkutlu, 2013), where the in situ density is somewhat higher than the density in the free-gas phase but significantly lower than that in the adsorption phase. In previous studies, CH_4 molecules are proposed to be mono-layered adsorption on the organic pore surface (Li and Jin, 2014; Dong et al., 2016). They proposed that the width of the adsorption layer formed by CH_4 molecules is 0.37 nm, being identical to the diameter of CH_4 molecule (Tian et al., 2017; Liu et al., 2018a,b). In this work, we also use the value of 0.37 nm as the width of the adsorption phase of CH_4 in nanopores.

3.3.4 Adsorbed CH_4 density in nanopores

By identifying the region of the adsorption phase in density profiles, the averaged adsorbed CH_4 density in the adsorption phase is calculated at the experimental temperature and pressure conditions. It is noted that the pore sizes of 2.23, 3.25, 4.20, and 6.21 nm dominates in shale samples #1, #2, #3, and #4, respectively (see Fig. 3.30). Thereby, the density is calculated in these four different pores with $\rho_{ave} = \int_a^b \rho_{ads}(z)dz/z_{ab}$ (ρ_{ave} represents the averaged CH_4 density in the adsorption phase; ρ_{ads} represents the in situ CH_4 density in the adsorbed layer; and z_{ab} represents the distance between points a and b, as depicted in Fig. 3.36). Fig. 3.37 presents the calculated averaged CH_4 density in the adsorption phase on the four shale samples. The averaged CH_4 density is highly related with temperature and pressure. Specifically, the averaged CH_4 density increases with increasing pressure, while it decreases with increasing temperature. Moreover, the averaged adsorbed CH_4 density varies for different shale samples. The difference in the pore size, TOC content, and the content of clay minerals is expected to contribute to the different adsorbed CH_4 density in the four shale samples. Therefore, it is not physically reasonable to employ a constant adsorbed CH_4 density to calculate the absolute adsorption in previous studies (Dubinin, 1960; Gensterblum et al., 2009; Ambrose et al., 2012; Jin et al., 2013; Wang et al., 2016a,b; Xiong et al., 2017).

3.3.5 Validation of the SLD model

To validate the CH_4–CH_4 force parameters used in the SLD model, the average CH_4 density in the free-gas phase is calculated in a 4.20-nm pore at 333.15 K; this density is then compared with that calculated from GCMC simulations and with the bulk CH_4 density obtained from NIST Chemistry WebBook (Lemmon et al., 2009). Fig. 3.38 shows the

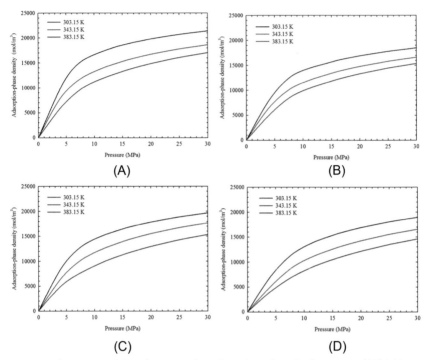

Fig. 3.37 The averaged CH_4 density in the adsorption phase in the pores of (A) 2.23 nm, (B) 3.25 nm, (C) 4.20 nm, and (D) 6.21 nm, respectively, at different temperature and pressure conditions.

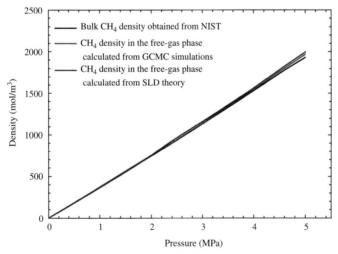

Fig. 3.38 Comparisons of CH_4 density in the free-gas phase in a 4.20-nm pore at 333.15 K among SLD theory, GCMC simulations, and the bulk CH_4 density obtained from the NIST Chemistry WebBook (Lemmon et al., 2009). The density values calculated from GCMC simulations are referred from our previous work (Liu et al., 2018a,b).

comparison results. It notes that the results calculated from GCMC simulations are cited from our previous work (Liu et al., 2018a,b). The average CH_4 density in the free-gas phase is calculated with $\rho_f = \int_c^{c'} \rho(z)dz / z_{cc'}$ (ρ_f represents the average density in the free-gas phase region; ρ represents the in situ density in the free-gas phase region; and $z_{cc'}$ represents the distance between points c and c') (as shown in Fig. 3.29). As shown in Fig. 3.38, compared to GCMC simulations, SLD theory has a relatively higher deviation from the NIST data, especially at higher pressure conditions. Overall, SLD model generally agrees well with the results from NIST data and GCMC simulations. It validates the reliability of the CH_4–CH_4 force parameters used in SLD theory.

From a molecular dynamic perspective, the free-gas phase of CH_4 in nanopores is mainly controlled by the CH_4–CH_4 force, while the adsorption behavior of CH_4 on pore surface is also controlled by the CH_4-pore surface interactions (Jiang and Lin, 2018). In our previous study, GCMC simulations have been successfully applied to predict the adsorption-phase density (Liu et al., 2018a,b). To validate the CH_4-pore surface parameters used in our SLD theory, we compare the calculated adsorption-phase density to that obtained from GCMC simulations. Fig. 3.39 shows the averaged CH_4 density in the adsorption phase in a 4.20-nm pore at different

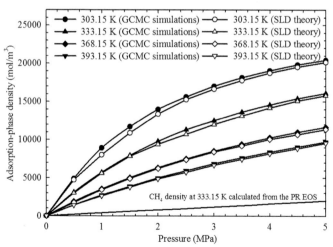

Fig. 3.39 The averaged CH_4 density in the adsorption phase in a 4.20-nm pore at different temperature and pressure conditions. We also calculate the CH_4 density at 333.15 K from the PR-EOS (1978) for comparison. The density values calculated from GCMC simulations are referred from our previous work (Liu et al., 2018a,b).

temperatures and pressures calculated from SLD theory and GCMC simulations. The adsorbed CH_4 density predicted by SLD theory slightly deviate from that obtained from GCMC simulations; such deviation enlarges as temperature decreases. The maximum deviation is calculated as -6.20%; it indicates that SLD theory is in a reasonable agreement with the GCMC simulations, which validates the CH_4-pore surface parameters used in the SLD theory in describing adsorption behavior of CH_4 on pore surface.

3.3.6 Absolute adsorption isotherms of CH₄ obtained from SLD model

Within the framework of SLD theory, fluid-pore surface interaction energy, ε_{fs}/k, and covolume correction parameter, b_{ads}, are two parameters that should be adjusted to match the measured excess CH_4 adsorption. Based on previous studies, the covolume factor is proposed to be reliable in the range of -0.3 to 0.3 (Mohammad et al., 2011; Pang et al., 2018). Besides the two parameters, i.e., fluid-pore surface interaction energy and the covolume correction parameter, some physical properties of the shale samples, i.e., the dominant pore size and BET surface area, are also required to input into the SLD model. As shown in Fig. 3.30, the dominant pore size for each shale sample can be obtained, while the BET surface area of the shale samples is measured from the N_2 adsorption/desorption tests (see Table 3.5).

Table 3.5 Parameters used in the SLD theory for fitting the excess CH_4 adsorption isotherms on each shale sample.

Shale sample	L (nm)	ε_{fs}/k (K)	b_{ads}
#1		82.0	0.051
	4.12	77.5	0.080
		74.2	0.055
#2		69.5	0.150
	2.23	67.5	0.155
		66.5	0.130
#3		65.3	0.060
	3.25	62.5	0.050
		56.2	0.045
#4		55.2	0.050
	6.21	52.6	0.060
		50.2	0.060

Table 3.6 TOC contents and BET surface areas of the two shale samples.

Shale sample ID	TOC content (wt%)	Porosity (%)	BET surface area (m²/g)
#1	2.85	3.26	2.98
#2	0.79	2.13	1.62

Table 3.6 lists the regression parameters for each shale sample. The curve-fitting results for the four shale samples are presented in Fig. 3.40. It indicates that SLD theory can properly represent the excess CH_4 adsorption on the four shale samples. Using Eq. (3.18), the excess adsorption is consequently converted into absolute adsorption isotherms, as depicted in Fig. 3.40. The directly measured excess adsorption is always lower than the absolute CH_4 adsorption. The difference between the excess and absolute CH_4 adsorption enlarges at lower temperature but higher-pressure conditions. It highlights the significance of using accurate adsorbed CH_4 density to obtain the absolute CH_4 adsorption, which is important for the shale gas-in-place estimation.

In previous work (Liu et al., 2018a,b), GCMC simulation method is also used to calculate the adsorbed CH_4 density and obtain the absolute CH_4 adsorption on shale. However, GCMC simulation method is quite computationally expensive. Based on the obtained results, SLD theory can nicely capture the adsorption-phase density of CH_4 on shale and thus can accurately determine the absolute adsorption. Compared to sophisticated molecular simulations (i.e., GCMC simulations), SLD theory is a more efficient method in obtaining the adsorbed CH_4 density. Thereby, it can be an alternative to other available engineering models.

In the future work, SLD theory is suggested to investigate fluid phase behavior in confined nanopores, which provides critical insights into the phase-behavior modeling in unconventional shale reservoirs. SLD theory can possibly be coupled into reservoir simulations by obtaining the phase behavior of confined fluids of various composition, pressure, and temperature. Pore-size distribution can possibly affect the adsorption behavior modeling. However, the recent works (Liu et al., 2018a,b) obtain the adsorption-phase density only in single nanopores. SLD theory is recommended to study adsorption behavior in nanoporous media, which is more practical for shale reservoir modeling. In addition, pore structures in shale may not only be slit-shaped, but also include cylindrical and

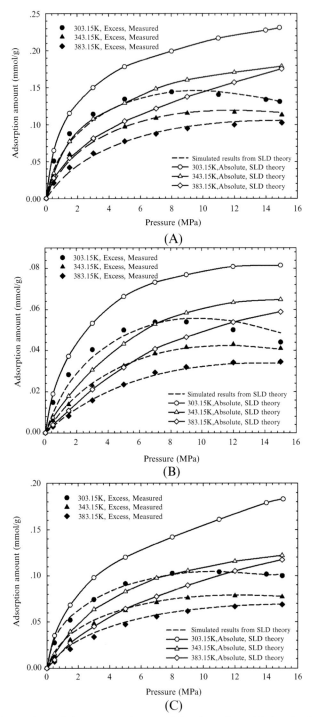

Fig. 3.40 Comparison of the measured excess CH$_4$ adsorption with the absolute adsorption obtained from the SLD theory on the four shale samples (A) #1, (B) #2, (C) #3, and
(Continued)

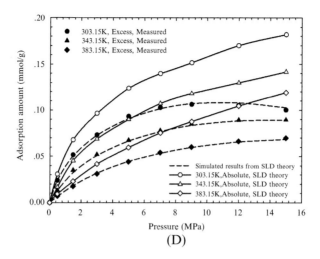

Fig. 3.40, cont'd (D) #4. It is noted that we also show the curve-fitting results in this figure.

ink–bottle shapes (de Boer and Lippens, 1964; Sing et al., 2008). Adsorption isotherms of hydrocarbons in different shaped pores are different from those in the slit-shaped pores. For example, within the ink–bottle model, depending on the pore-diameter ratio of "ink" and "bottle," nitrogen adsorption/desorption isotherms may behave differently as the result of pore-blocking and cavitation effects (Keffer et al., 1996; Fan et al., 2011; Klomkliang et al., 2013; Dobrzanski et al., 2018). Future work is suggested to explore the effect of pore-size distribution and pore geometry on the adsorption behavior of pure hydrocarbon or hydrocarbon mixtures in confined space.

As for some more complex molecules, such as C_2H_6, C_3H_8, etc., there may be a second adsorption layer formed in nanopores. However, SLD model cannot describe the second adsorption layer for these molecules. In the future works, it is necessary to revise the SLD model to investigate the confined fluid properties of the complex molecules by considering more complex fluid-pore wall interactions as well as the chemical potential calculations. Besides CH_4, shale gas also contains secondary components, such as C_2H_6, C_3H_8 as well as N_2 etc. Desorption generally dominate the whole shale gas production; it is thereby necessary to investigate the desorption behavior of shale gas mixtures during shale gas exploration (Dasani et al., 2017).

3.4 Determination of the absolute adsorption isotherms of CH_4 on shale with low-field nuclear magnetic resonance

Shale gas has been an increasingly important energy resource in recent years, especially in America. Multicomponents are generally contained in shale gas, while CH_4 dominates in most shale gas reservoirs. Shale CH_4 stored in shale gas reservoirs can be in the following states, i.e., adsorbed gas on rock surface, compressed gas in natural and hydraulic fractures, compressed gas in matrix porosity, and absorbed gas in organic matter (Clarkson and Haghshenas, 2013). Shale reservoirs are usually organic-rich, resulting in a large amount of CH_4 in the absorbed state (Ross and Bustin, 2009). Thereof, to obtain the initial gas storage in shale reservoirs, it is crucial to accurately calculate the amount of adsorbed gas, which is, however, the basis for disclosure of gas reserves and is also important for reservoir-engineering analysis, such as gas-production forecast (Ambrose et al., 2012).

Thermogravimetric analysis (TGA) technique and volumetric method are two main experimental approaches in obtaining the adsorption isotherms. However, the adsorption isotherms directly measured by TGA technique and volumetric method are excess adsorption isotherms, which neglect the adsorbed-phase volume and thereby underestimates the total adsorption amount. Absolute adsorption is defined as the adsorbed amount of gas in the adsorbed state (Heller and Zoback, 2014). With the knowledge of the absolute adsorption, the exact amount of the adsorbed gas in shale reservoir can be accurately determined. The adsorption phase density is generally used to obtain the absolute adsorption isotherms by calibrating the measured excess adsorption isotherms. The adsorption phase density is defined as the density of the adsorbed CH_4 on the pore surface in shale samples. Such density is difficult to be measured in lab; to our knowledge, few efforts are dedicated to quantifying the density of the adsorption phase. In the previous studies, the adsorption phase density is normally represented by constant values (Dubinin, 1960; Menon, 1968; Tsai et al., 1985; Wang et al., 2016a,b). They persuaded that the adsorption phase correlates with the van der Waals constant b or is equal to the liquid adsorbate density. However, it has been found that the density of the adsorption phase is a function of the system pressure, temperature, and pore size (Ambrose et al., 2012). To accurately determine the absolute adsorption isotherms, it is, thereby, of critical importance to precisely capture the adsorption phase density. From a nanopore-scale perspective, molecular simulations, such as grand canonical Monte Carlo (GCMC) simulations, can faithfully capture the properties of the adsorption phase over a wide pressure and temperature range due to the consideration of fluid/surface interactions. However, the GCMC simulation is expensive in obtaining the adsorption phased density due to its time-consuming computation. Furthermore, the current method in obtaining the absolute adsorption isotherms is quite complex; that is, the excess adsorption isotherms have to be initially measured and then one have to calculate the absolute adsorption isotherms using the molecular simulations.

Low-field NMR technique has been widely used to investigate the distributions of hydrogen-containing fluids existing in porous media (Coates et al., 1999). In oil industry, this technique can be applied to study the porosity, the pore-size distribution, and measure the permeability of reservoir rocks etc. (Yao et al., 2010) The NMR technique has also been used in coal industry to investigate the ^1H NMR relaxation of CH_4 in bulk (Gerritsma et al., 1971; Akkurt et al., 1996). By knowing that CH_4 can exist in shale rocks in the adsorbed state, many studies have been conducted to reveal the adsorbed CH_4 in coal samples using the NMR technique (Alexeev et al., 2004; Guo et al., 2007; Yao et al., 2014). By measuring the T_2 spectrum of CH_4 in coals, they proposed that CH_4 can adsorb or dissolve in solid coals besides of the free-gas state in the bulk or closed pore or fracture system (Alexeev et al., 2004; Guo et al., 2007; Yao et al., 2014). Yao et al. (2014) measured three types of T_2 peaks for CH_4 in coal samples; they suggested that these peaks correspond to the bulk CH_4 in the free space between the coal particles and within the sample cell, free-state CH_4 in large pores or fractures, and the adsorbed CH_4 on the inner pore surface. Based on the measurement, they obtained the adsorption isotherms of CH_4 on coal samples. Until now, the NMR technique is scarcely employed in the shale industry to obtain the adsorption isotherms of CH_4 on shale samples.

The objectives of this section are to propose a pragmatic method to determine the absolute adsorption isotherms of CH_4 using NMR technique. In order to validate this method, we measure the excess adsorption isotherms of CH_4 on the same shale samples, and then determine the absolute adsorption isotherms with the GCMC simulations. The absolute adsorption isotherms obtained from two approaches are compared and the capability of the NMR technique in obtaining the absolute adsorption isotherms is evaluated. Within our knowledge, it is the first time that the NMR technique is applied to determine the absolute adsorption isotherms of CH_4 on shale samples.

3.4.1 Characterization of the shale samples

A combustion elemental analyzer is used to measure the TOC content of the two shale samples. In the first step, the organic carbon in shale samples is sparged with oxygen by forming carbon dioxide. The nondispersive infrared detector is then used to detect the total amount of the carbon dioxide. Thereby, the TOC content of the two shale samples can be determined, as shown in Table 3.6. Shale sample #1 has a TOC content of 2.85 wt%, which is 3.61 times of that in shale sample #2.

The surface morphology of the shale samples is characterized by the Hitachi TM-300 SEM setup. Prior to the scanning, a golden film is coated on the shale surface with a thickness of 10 nm to improve the conductivity. Fig. 3.41 presents the digital FE-SEM images taken on the two shale samples. Bunches of pores are observed coexisting in both shale samples. As marked in Fig. 3.41, the energy-dispersive X-ray spectroscopy (EDX) analysis is further conducted on the chosen points "a" and "b" in shale samples #1 and #2, respectively. Fig. 3.42 shows the results obtained from the EDX test. At the marked points, a high concentration of carbon element is observed, suggesting that the organic matter should exist in these sites, i.e., kerogen.

The N_2 adsorption/desorption tests are conducted to obtain the pore-size distributions of the two shale samples by using the Autosorb iQ-Chemiadsorption and Physi-adsorption Gas Adsorption Analyzer (Quantachrome Instruments, United States). Fig. 3.43 presents the pore-size distributions of the two shale samples. The pores in the shale sample #1 exhibit pore sizes mainly around 6.0 nm, while the pore size in the shale sample #2 is

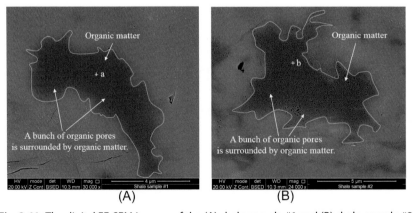

Fig. 3.41 The digital FE-SEM images of the (A) shale sample #1 and (B) shale sample #2.

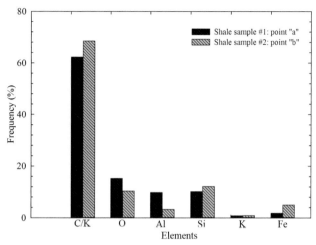

Fig. 3.42 Energy-dispersive X-ray spectroscopy (EDX) analysis results for points "*a*" and "*b*" in shale samples #1 and #2, respectively.

mainly around 10.0 nm. The BET surface area is obtained for the two shale samples as well based on the BET equation (Brunauer et al., 1938), as shown in Table 3.6. The BET surface area for shale sample #1 is 2.98 m^2/g, which is much higher than that of the shale sample #2. Based on the N_2 adsorption/desorption tests, shale sample #1 has a higher porosity than that of shale sample #2, which may correlate with the higher TOC content and higher surface area in the shale #1.

3.4.2 Measurements of the excess adsorption of CH_4 on shale samples

The excess adsorption isotherms of CH_4 are measured using a thermogravimetric analyzer (TGA) (IEA–100B, Hiden Isochema Ltd., United Kingdom). The key component of TGA is a magnetic suspension balance with a 1.0 μg accuracy in weight measurement. The test pressures are set up to 12.0 MPa, while the test temperature is set at 298.15 K.

TGA uses the thermogravimetric analysis method to obtain the excess adsorption amount by knowing the weight change of the shale sample. A brief procedure for obtaining the excess adsorption amount is given as follows. First, the weight (M_{sc}) and volume (V_{sc}) of the empty sample container are measured at a preset temperature. Then, the weight (M_s) and volume (V_s) of the shale sample are measured by placing the shale sample into the adsorption chamber. Subsequently, after being vacuumed for 6 h, the chamber is

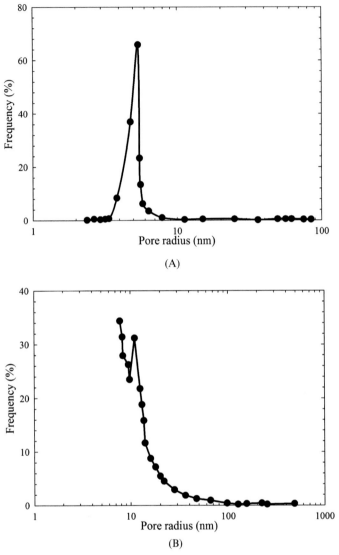

Fig. 3.43 Pore-size distributions of (A) shale sample #1 and (B) shale sample #2 as obtained from the N_2 adsorption/desorption test.

filled with CH_4. The pressure of the sample chamber is then increased by a stepwise manner to the operating pressure. When reaching stabilization, the apparent weight (ΔM) is calculated by (Wang et al., 2015),

$$\Delta M = M_{sc} + M_s + M_a - (V_{sc} + V_s + V_a)\rho \tag{3.35}$$

where M_a is the mass of gas adsorbed on shale sample, which is the so called absolute adsorption; V_a is the adsorbed gas volume; ρ is the bulk gas density obtained from National Institute of Standards and Technology's (NIST) Chemistry WebBook (Lemmon et al., 2009).

Then, reorganizing Eq. (3.35) leads to the expression of M_a,

$$M_a = \Delta M - M_{sc} - M_s + (V_{sc} + V_s + V_a)\rho \qquad (3.36)$$

Consequently, the excess adsorbed mass (M_{ex}) can be obtained,

$$M_{ex} = M_a - \rho V_a = \Delta M - M_{sc} - M_s + (V_{sc} + V_s)\rho \qquad (3.37)$$

Each test is repeated two times to make sure the measured results are reliable and reproducible. The maximum deviation between two consecutive runs is found to be less than $\pm 3.25\%$.

3.4.3 Absolute adsorption of CH_4 on shale samples

With the TGA method, V_a is obtained by measuring the void volume with helium (He). Note that as adsorption occurs, the void volume is actually decreasing due to the presence of the adsorbed phase. However, M_{ex} ignores the adsorbed-gas volume (see Eq. 3.37), and tends to underestimate the total adsorption mass (M_a). In this study, M_{ex} is corrected by the following expression to obtain the total adsorption mass (M_a), that is equal to the so called absolute adsorption mass (M_{abs}), by considering the adsorbed-phase density (ρ_a) (Menon, 1968; van der Sommen et al., 1955; Ozawa et al., 1976):

$$M_{abs} = \frac{M_{ex}}{1 - \dfrac{\rho}{\rho_a}} \qquad (3.38)$$

Based on Eq. (3.38), the absolute adsorption uptake is obtained by accurately calculating the adsorbed-phase density. Previously, the adsorbed-phase density was employed as a constant value which was generally obtained from the liquid density (Menon, 1968; Tsai et al., 1985; Wang et al., 2016a,b) or either calculated from the van der Waals constant b (Dubinin, 1960). However, it has been found that the adsorbed-phase density strongly correlates with the system pressure, temperature, and pore size (Ambrose et al., 2012). Molecular simulations can capture the properties of the adsorption phase by faithfully considering the pore surface/fluid interactions. In this work, the adsorbed-phase density is calculated using the GCMC simulations.

3.4.4 NMR test of CH_4's adsorption on shale samples

Fig. 3.44 shows the schematic diagram of the NMR setup used for analyzing the adsorption of CH_4 in shale samples. A syringe pump (ISCO-500D, Teledyne Isco, Lincoln, NE, United States) is used to inject CH_4 or He into the shale samples contained in a core holder (Niumag, Shanghai); such a core holder is made of nonmagnetic material PEEK which can withstand pressure and temperature up to 20.0 MPa and 80.0°C, respectively. This core holder is placed in a magnetic coil and the NMR test apparatus (Mini-MR, Niumag, China) is used to measure the T_2 spectrum of the shale sample at different pressure stages. The detailed parameters used in this study are given in Table 3.7.

Before NMR test, the whole gas–delivery system, including core holder and connecting tubes, is evacuated for 120 min using vacuum pump; then,

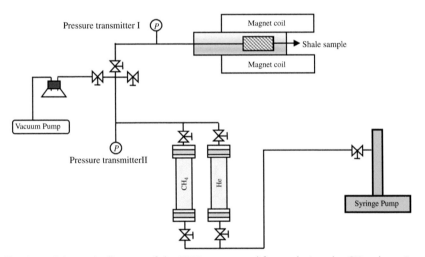

Fig. 3.44 Schematic diagram of the NMR setup used for analyzing the CH_4 adsorption on shale samples.

Table 3.7 The specific values of the parameters used in the NMR test.

Injection pressure (MPa)	0–2	2–4	4–6	6–8	8–10	10–12
Waiting time (s)	1.5	3.0	4.0	6.0	8.0	10.0
Echo spacing (ms)	0.3	0.3	0.3	0.3	0.3	0.3
Field intensity (T)	0.5	0.5	0.5	0.5	0.5	0.5
Number of train	64	64	64	64	64	64
Number of echo	18,000	18,000	18,000	18,000	18,000	18,000

He is injected into the shale samples to a certain pressure and the NMR test is applied to obtain the T_2 spectrum of the dry shale samples. Consequently, the whole system is evacuated again and then a certain amount of CH$_4$ is injected into the core holder to achieve a testing pressure. It is noted that the injected gas amount can be obtained with the knowledge of the injected gas volume and gas density at given pressures. The specific gas densities are read from National Institute of Standards and Technology's (NIST) Chemistry WebBook (Lemmon et al., 2009), while the injected volume can be obtained by the volume change in the transfer cylinder containing CH$_4$. The injected volume and the specific gas densities for different injecting pressures are given in Table 3.8. Following the gas injection, the NMR test is initiated to obtain the T_2 spectrum of CH$_4$ in shale samples. Then, after certain soaking time, CH$_4$ is released in a stepwise manner from another end of the core holder with the equilibrium pressure decreasing continuously. The measured T_2 spectrum is used to infer the gas states existing in shale samples. And, the variation in the amplitude of T_2 spectrum at different equilibrium pressures is used to obtain the amount of CH$_4$ adsorbed on the shale sample.

Table 3.8 The specific injected molar amount of CH$_4$ at 298.15 K and given pressures.

Sample ID	Mass (g)	Equilibrium pressure (MPa)	Injected gas volume (mL)	Gas density (kg/m^3)
#1	49.02	0.00	0.00	0.000
		1.50	2.41	10.077
		3.15	1.99	21.470
		5.23	1.73	36.967
		7.12	1.60	52.037
		9.32	1.48	70.187
		11.15	1.39	85.919
#2	48.53	0.00	0.00	0.000
		1.45	1.16	9.634
		3.23	1.12	22.095
		5.32	0.89	37.802
		7.25	0.92	53.092
		9.45	0.88	71.287
		11.35	0.88	87.596

The equilibrium pressure means the pressure at which the adsorption reaches equilibrium.

3.4.5 Nuclear magnetic resonance (NMR) technique

When certain atoms, e.g., hydrogen proton, are placed in a static magnetic field and exposed to a second oscillating magnetic field, the nuclear magnetic resonance will occur (Anatoly, 2013). In a porous media (e.g., shale samples), the T_2 spectrum of CH$_4$ in a magnetic field is influenced by the bulk relaxation, diffusion in magnetic gradients, and surface relaxation, as depicted by (Megawati et al., 2012),

$$\frac{1}{T_2} = \frac{1}{T_{2B}} + \frac{1}{T_{2D}} + \frac{1}{T_{2S}} \tag{3.39}$$

where T_{2B}, T_{2D}, and T_{2S} represent the transverse time due to bulk relaxation, diffusion in magnetic gradients, and surface relaxation, respectively, ms. In shale cores, the diffusion relaxation is small enough to be neglected (Bloembergen et al., 1948), while the surface relaxation is strongly related to the shale core's surface area, i.e., the ratio of the pore's surface area to the pore volume. Thereby, Eq. (3.39) becomes,

$$\frac{1}{T_2} = \frac{1}{T_{2B}} + \frac{1}{T_{2S}} = \frac{1}{T_{2B}} + \rho\frac{S}{V} \tag{3.40}$$

where ρ is the relaxation rate, μm/ms; S/V is the specific area of pore, 1/μm. It indicates that the relaxation of CH$_4$ in shale samples is determined by both the bulk properties and the surface effect.

3.4.6 Grand canonical Monte Carlo (GCMC) simulations

Within the grand canonical ensemble, the properties of the entire system, i.e., volume (V), temperature (T), and chemical potential (μ), are fixed. In the ensemble, the average number of molecules is determined by the chemical potential, since the number of molecules in the system fluctuates during the simulations.

In this model, different hydrocarbon molecules are simulated by the united atom model (Martin and Siepmann, 1998). The modified Buckingham exponential-6 intermolecular potential (Errington and Panagiotopoulos, 1999) is used to describe the nonbonded site-site interactions among functional groups on different molecules, as well as among functional groups belonging to the same molecule separated by more than three bonds. The pairwise interaction potential $U(r)$ for the nonbonded site-site interactions is expressed as (Errington and Panagiotopoulos, 1999),

$$U(r) = \begin{cases} \dfrac{\varepsilon}{1-\dfrac{6}{\alpha}}\left[\dfrac{6}{\alpha}\exp\left(\alpha\left[1-\dfrac{r}{r_m}\right]\right)-\left(\dfrac{r_m}{r}\right)^6\right], r > r_{max} \\ \infty, \ r < r_{max} \end{cases} \quad (3.41)$$

where r is the interparticle separation distance, r_m is the radial distance at which $U(r)$ reaches a minimum, and the cutoff distance r_{max} is the smallest radial distance for which $dU(r)/dr=0$ (Singh et al., 2009). Since the original Buckingham exponential-6 potential can be negative at very short distances, the cutoff distance is thus defined to avoid negative potentials (Errington and Panagiotopoulos, 1999). The radial distance at which $U(r)=0$ is defined as σ. The values of the exponential-6 parameters ε, σ, and α are 160.3 K, 0.373 nm, 15, respectively, for CH_4 (Singh et al., 2009). The combining rules are applied to determine the cross parameters as (Singh et al., 2009),

$$\sigma_{ij} = \frac{1}{2}\left(\sigma_i + \sigma_j\right) \quad (3.42)$$

$$\varepsilon_{ij} = \left(\varepsilon_i\varepsilon_j\right)^{1/2} \quad (3.43)$$

$$\alpha_{ij} = \left(\alpha_i\alpha_j\right)^{1/2} \quad (3.44)$$

The bond bending potential $U_{bend}(\theta)$ is calculated by (van der Ploeg and Berendsen, 1982),

$$U_{bend}(\theta) = \frac{K_\theta}{2}\left(\theta - \theta_{eq}\right)^2 \quad (3.45)$$

where parameter K_θ is equal to 62,500 K/rad^2, θ is the bond angle from equilibrium, and θ_{eq} is the equilibrium bond angle (114 degrees).

It has been found that CH_4 can heavily adsorb on the organic-pore surface due to the higher organic carbon content in shale samples (Kim et al., 2017a,b). Thereby, slit geometry with smooth and structureless carbon surfaces is selected to represent the nanopores. The fluid-pore surface interactions φ_{wf} are described by the 10-4-3 Steele potentials as (Steele, 1973),

$$\varphi_{wf}(z) = 2\pi\rho_w\varepsilon_{wf}\sigma_{wf}^2\Delta\left[\frac{2}{5}\left(\frac{\sigma_{wf}}{z}\right)^{10} - \left(\frac{\sigma_{wf}}{z}\right)^4 - \frac{\sigma_{wf}^4}{3\Delta(0.61\Delta + z)^3}\right] \quad (3.46)$$

where z is the distance of the fluid particle from the pore surface, ρ_{wf} is the density of carbon atom per unit surface area of the graphite layer (114 nm^{-3}), the molecular parameters of an atom in the graphite layer are $\varepsilon_{wf} = 28\,K$, and $\sigma_{wf} = 0.3345\,nm$ (Do and Do, 2003), and Δ is the spacing between two adjacent graphene layers (0.335 nm), respectively. The external potential ψ in a slit pore is given as (Steele, 1973),

$$\psi(z) = \varphi_{wf}(z) + \varphi_{wf}(W - z) \tag{3.47}$$

where W is the size of the slit pore.

In each Monte Carlo (MC) cycle, a trial random displacement is applied to all CH_4 molecules; with equal probability, a CH_4 molecule is randomly removed from or inserted into the simulations box depending on the chemical potential of CH_4. The Widom insertion method (Widom, 1963) is applied to obtain the chemical potentials of the bulk CH_4 molecules in the canonical ensemble. The PR-EOS (Peng and Robinsion, 1976) is applied to calculate the bulk densities of CH_4 at given pressures and temperatures. The MC moves are implemented by using the Metropolis algorithm (Metropolis et al., 1953). During the simulations, 0.1 million of MC cycles per each adsorbate molecule is required to reach an equilibrium state, while 0.5 million of MC cycles per adsorbate molecule is required to sample the density profiles.

The average density (ρ_{ave}) of CH_4 in carbon-slit pores is expressed as,

$$\rho_{ave} = \frac{\langle N_i \rangle M_i}{V N_A} \tag{3.48}$$

where $\langle N_i \rangle$ is the ensemble averaged number of CH_4 molecules in nanopores, V is the volume, M is molecular weight of CH_4, and N_A is Avogadro constant.

3.4.7 Excess and absolute adsorption isotherms of CH_4 on shale

3.4.7.1 Excess adsorption isotherms

With TGA, the excess adsorption isotherms of CH_4 are measured on two shale samples at 298.15 K. Fig. 3.45 presents the measured excess adsorption isotherms. The excess adsorption strongly correlates with the system pressure: it increases as pressure increases and then levels off after certain pressure. Comparing with shale sample #2, CH_4 exhibits a higher excess adsorption capacity on shale sample #1. It has been found that the adsorption capacity is related with TOC content and surface area of shale samples (Kim et al., 2017a,b). Thereof, such higher adsorption on shale sample #1 may be

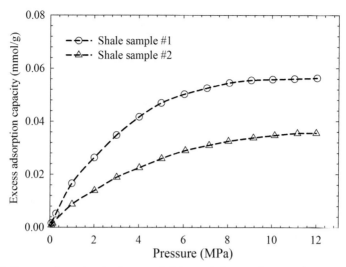

Fig. 3.45 Excess adsorption isotherms of CH_4 on shale sample #1 and #2 at 298.15 K.

caused by the higher TOC content (2.85 wt%) and larger BET surface area (2.98 m^2/g) than that of shale sample #2 which has a TOC content of 0.79 wt% and a BET surface area of 1.62 m^2/g.

3.4.7.2 Adsorption-phase density

To obtain the absolute adsorption isotherms, GCMC simulations are applied to investigate the CH_4 distributions in nanopores. Based on the pore-size distributions of the two studied shale samples (see Fig. 3.43), two pore sizes are selected, i.e., 6.0 and 10.0 nm corresponding to the dominant pore size of shale sample #1 and #2, respectively. Fig. 3.46 presents the density distributions of CH_4 in 6.0- and 10.0-nm pores at 298.15 K and at different system pressures. At all bulk pressure conditions, CH_4 molecules can form one strong adsorption layer and the density of the adsorption-layer is significantly higher than the density at the pore center, while the density in the pore center approaches the bulk density obtained from National Institute of Standards and Technology's (NIST) Chemistry WebBook (Lemmon et al., 2009). When pressure is larger than 2.0 MPa, a second weak adsorption layer can form in the location adjacent to the first adsorption layer. As pressure increases, the density in the adsorption phase is found to increase, which indicates that the density of the adsorption phase correlates with the system pressure. As depicted in Fig. 3.46, the region locating between a (or a') and b (or b') is defined as the adsorption phase by assuming the CH_4 adsorption is

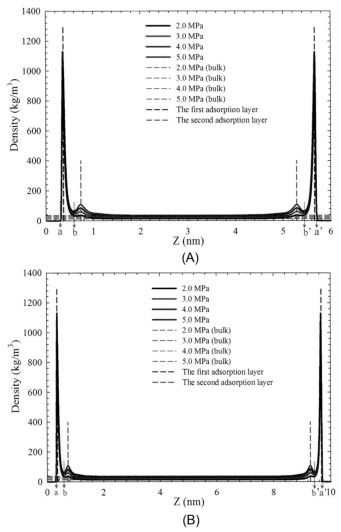

Fig. 3.46 Density profiles of CH_4 in the carbon-slit pore of (a) 6.0 and (b) 10.0 nm at 298.15 K and different pressures.

single-layered adsorption which is in line with the previous study (Li et al., 2014). The region locating between b and b' is defined as the free-gas phase. Point b (or b') is selected as the saddle point between the first adsorption layer and the second weak adsorption layer. We observe that the width of the adsorption phase is almost the same at different system pressures; the width of the adsorption phase of CH$_4$ (ab), around 0.37 nm, is similar to the diameter of CH$_4$ molecule.

Based on the density profiles of CH$_4$ in nanopores, we thereupon calculate the average density of the adsorption phase for CH$_4$ by,

$$\rho_{ave} = \int_a^b \rho_{ads}(z)\,dz/z_{ab} \tag{3.49}$$

where ρ_{ave} is the average density of the adsorption phase, ρ_{ads} is the in situ density of the adsorption phase in nanopores, and z_{ab} is the distance between a and b (see Fig. 3.47).

The density of the adsorption phase of CH$_4$ in carbon-slit pores of 6.0 and 10.0 nm is calculated at different pressures and at 298.15 and 333.15 K, as shown in Fig. 3.47. In this figure, the average density of the adsorption phase of CH$_4$ in 6.0-nm pore is identical to that in 10.0-nm pore. It implies that for both pore sizes, a change in the pore size will not affect the configuration of the adsorption layers formed by the CH$_4$ molecules. Furthermore, the average density of the adsorption phase strongly correlates with the system pressure and temperature: it increases as the system pressure increases, while it decreases as the temperature increases. It should be noted that the constant liquid CH$_4$ density has been extensively used as the

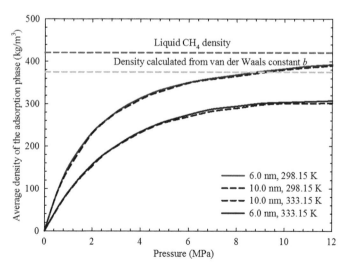

Fig. 3.47 Average density of the adsorption phase of CH$_4$ confined in the carbon-slit pore of 6.0 nm and 10.0 nm at different temperatures and pressures. The constant density of liquid CH$_4$ and the density calculated from van der Waals constant b are also shown in this figure.

density of the adsorption phase to obtain the absolute adsorption isotherms (Wang et al., 2016a,b) or fit empirical models to the adsorption isotherms (Weniger et al., 2010; Gasparik et al., 2012; Rexer et al., 2013). Therefore, it may not be appropriate to use a constant density value to represent the density of the adsorption phase (Heller and Zoback, 2014; Wang et al., 2016a,b).

The liquid CH_4 density has been extensively used as the density of the adsorption phase to obtain the absolute adsorption isotherms (Wang et al., 2016a,b). The constant value of $421 \, kg/m^3$ is mostly used. The constant density of CH_4 based on the van der Waals constant b is also heavily used to represent the density of the adsorption phase (Menon, 1968; Rexer et al., 2013), i.e., $1/b$. Fig. 3.47 also shows the density of liquid CH_4, $421 \, kg/m^3$ and another constant density of CH_4 calculated from the van der Waals constant b. It is clear that, as for either CH_4, the density of the adsorption phase should be a variable which depends on the in situ temperature/pressure, rather than a constant value.

3.4.7.3 Absolute adsorption isotherms

The measured excess adsorption isotherms are converted to the absolute adsorption isotherms using the averaged density of the adsorption phase which are calculated from GCMC simulations. In Fig. 3.48, the excess

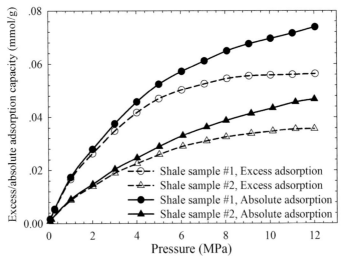

Fig. 3.48 Comparison between the excess and the absolute adsorption isotherms of CH_4 on shale samples #1 and #2.

adsorption isotherms of CH_4 are compared with the corresponding absolute adsorption isotherms. The absolute adsorption of CH_4 is found to be always higher than the directly measured excess adsorption. A relatively larger deviation is found to exist between the absolute adsorption isotherms and the excess adsorption isotherms at relatively higher pressures, which highlights the importance of using accurate density of the adsorption phase to obtain accurate absolute adsorption isotherms for CH_4 at the high pressure region.

3.4.8 T_2 spectrum of CH_4 in shale samples

Fig. 3.49 presents the T_2 spectrum of CH_4 in shale samples #1 and #2. Generally, three distinct T_2 peaks can be observed and the system pressure results in an overall increase of the T_2 peaks. Based on Eq. (3.40), the shale surface is inferred to enhance such three types of relaxation of CH_4 in the shale-filled system, i.e., 0.1–13, 13–280, and 280–2500 ms. As depicted in Fig. 3.49, the dry shale samples show almost zero T_2 signal before injecting CH_4, which is mainly because that the shale samples were dried adequately at a high temperature, resulting in only tiny amount of residual water left in the shale samples. Yao et al. (2014) measured the CH_4 adsorption on dry coal using a low-field NMR relaxation method; they suggested that CH_4 can exist in coals in three-phase states: (a) CH_4 on the inner coal pore surfaces; (b) free-state CH_4 in pores and fractures; and (c) bulk CH_4 in the free space between the coal particles. According to Eq. (3.40), smaller pores have high values of S/V; protons in such pores relax faster than those in larger pores (Yao et al., 2014). Therefore, in this work, the T_2 peak within the range of 0.1–13 ms is inferred to represent CH_4 adsorbed on the shale surface, while the T_2 peak within the range of 13–280 ms is caused by free-state CH_4 in pores and fractures. The T_2 spectrum in the range of 280–2500 ms corresponds to the bulk CH_4 in the free space between the shale particles.

As depicted in Fig. 3.49, the amplitude of T_2 spectrum increases as system pressure increases. We calculate the integrated amplitudes of the adsorbed CH_4 on the pore surface (i.e., 0.1–13 ms), the free-state CH_4 in pores (i.e., 13–280 ms), and the bulk CH_4 (i.e., 280–2500 ms) at different pressures. With the knowledge of the injected gas volume and gas density obtained from NIST (Lemmon et al., 2009) (see Table 3.8), the total amount of injected CH_4 at specific pressures can be calculated, as shown in Table 3.9; the specific adsorbed amount of CH_4 (i.e., 0.1–13 ms), the amount of CH_4 in the free-state, and the amount of CH_4 in bulk are then obtained with the system pressure (as depicted by **a**, **b**, and **c** in Fig. 3.49, respectively). The

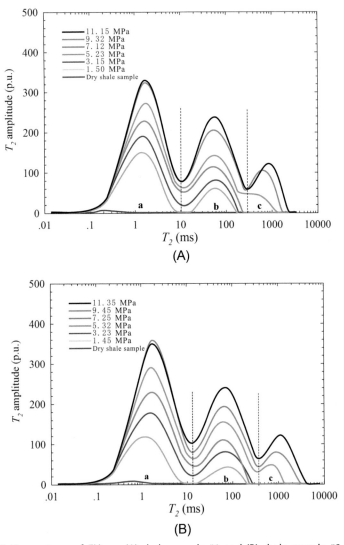

Fig. 3.49 T_2 spectrum of CH_4 on (A) shale sample #1 and (B) shale sample #2.

corresponding gas amount is depicted as a function of pressure in Fig. 3.50. It can be found that the "bulk CH_4" and the "free-state CH_4 in pores" is linearly proportional to the system pressure, while the "adsorbed CH_4 on the pore surface" increases as the pressure increases in the first place, then reaches a maximum value; this observation is in line with the previous studies by Yao et al. (2014).

Table 3.9 The specific amount of "adsorbed CH_4," "free-state CH_4," and "bulk CH_4" for given pressures.

Sample	Amplitude (%)			Total mole of injected CH_4 (mmol/g)	Specific mole of CH_4 (mmol/g)			Equilibrium pressure (MPa)
	a	b	c		a	b	c	
#1	71.2	28.8	0.0	0.0309	0.0220	0.0089	0.0000	1.50
	68.9	31.1	0.0	0.0544	0.0375	0.0169	0.0000	3.15
	63.6	36.4	0.0	0.0811	0.0516	0.0295	0.0000	5.23
	55.6	32.8	11.6	0.1061	0.0590	0.03480	0.0123	7.12
	49.6	35.2	15.2	0.1319	0.0654	0.0464	0.0201	9.32
	45.5	36.8	17.7	0.1523	0.0693	0.0561	0.0270	11.15
#2	70.1	29.9	0.0	0.0143	0.0100	0.0043	0.0000	1.45
	67.2	32.8	0.0	0.0315	0.0211	0.0103	0.0000	3.23
	68.6	31.4	0.0	0.0425	0.0292	0.0134	0.0000	5.32
	56.1	35.6	8.3	0.0618	0.0347	0.0220	0.0051	7.25
	50.3	36.2	13.5	0.0800	0.0401	0.0288	0.0108	9.45
	43.1	38.6	18.3	0.0985	0.0424	0.0380	0.0180	11.35

As shown in Fig. 3.50, the bulk CH_4 is zero at lower pressures. As depicted in Fig. 3.49, the region *c* is nearly absent at low pressures (<5.23 MPa for shale sample #1 and <5.32 MPa for shale sample #2). It is probably due to the fact that the volume concentration of "bulk CH_4" in the free space between the shale particles at low pressures is below the detection limit of the NMR setup. Since shale has a strong affinity to CH_4, CH_4 prefers to adsorb on the shale surface at low pressures. That is, at low pressures, the adsorption is the dominant mechanism. When pressure is high (>5.23 MPa for shale sample #1 and >5.32 MPa for shale sample #2), three peaks appear in the T_2 spectrum, as shown in Fig. 3.49, and the *b* and *c* peaks increases as the pressure increases. Above this pressure, the volume concentration of CH_4 filling the free space between the shale particles is sufficiently high to be detected by the NMR setup.

3.4.9 Comparison of the absolute adsorption isotherms from two approaches

Fig. 3.51 compares the absolute adsorption isotherms calibrated from the measured excess values with the absolute adsorption isotherms measured from the NMR method. The corresponding excess adsorption isotherms are also depicted in this figure. Interestingly, the absolute isothermal curves obtained from two approaches are in very good agreement, while the excess

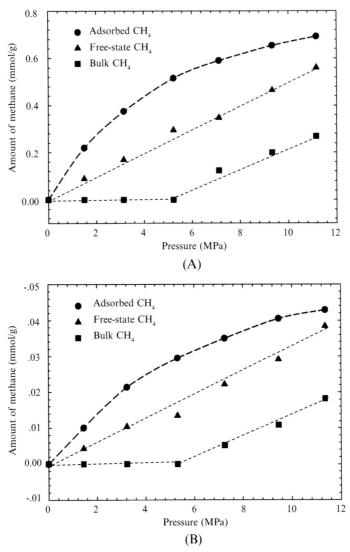

Fig. 3.50 The specific amount of "adsorbed CH_4," "free-state CH_4," and "bulk CH_4" on (A) shale sample #1 and (B) shale sample #2 as a function of pressure.

adsorption underestimates the actual adsorption capacity on shale samples (i.e., the absolute adsorption capacity). Compared with the calibrated absolute adsorption, the absolute adsorption obtained from the NMR method has an averaged deviation of −2.56%; that is, the NMR method slightly underestimates the absolute adsorption capacity calibrated by the adsorption phase density. It is possibly because that the NMR setup cannot detect the

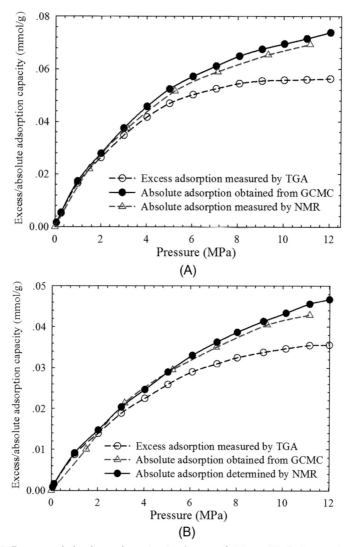

Fig. 3.51 Excess and absolute adsorption isotherms of CH_4 on (A) shale sample #1 and (B) shale sample #2.

CH_4 relaxing in less than $0.1\,ms$. Another possible reason may be caused by the dissolved CH_4 in shale matrix, which, however, also cannot be detected by such low-field NMR setup. Based on the comparison results, the low-field NMR technique provides an alternating method that can be used to obtain the absolute adsorption capacity other than by means of the complex molecular simulations.

Appendix 1: Position-dependent equation of state parameter [$a_{ads}(z)$]

When $1.0 \leq L/\sigma_{ff} \leq 1.5$, the mean field model will break down. Thereby, if, $L/\sigma_{ff} = 1.5$.

$$a_{ads}(z) = \frac{3}{16} \qquad (A1)$$

When, $1.5 \leq L/\sigma_{ff} \leq 2.0$ $a_{ads}(z)$ is given by,

$$\frac{a_{ads}(z)}{a(T)} = \frac{3}{8}\left(\frac{L}{\sigma_{ff}} - 1\right) \qquad (A2)$$

When $2.0 \leq L/\sigma_{ff} \leq 3.0$, $a_{ads}(z)$ is given by,

$$\frac{a_{ads}(z)}{a(T)} = \frac{3}{8}\left(\frac{z}{\sigma_{ff}} + \frac{5}{6} - \frac{1}{3}\left(\frac{L-z}{\sigma_{ff}} - \frac{1}{2}\right)^{-3}\right) \text{ for } 0.5 \leq \frac{z}{\sigma_{ff}} \leq \frac{L}{\sigma_{ff}} - 1.5 \quad (A3)$$

$$\frac{a_{ads}(z)}{a(T)} = \frac{3}{8}\left(\frac{z}{\sigma_{ff}} - 1\right) \text{ for } \frac{L}{\sigma_{ff}} - 1.5 \leq \frac{z}{\sigma_{ff}} \leq 1.5 \qquad (A4)$$

$$\frac{a_{ads}(z)}{a(T)} = \frac{3}{8}\left(\frac{L-z}{\sigma_{ff}} + \frac{5}{6} - \frac{1}{3}\left(\frac{z}{\sigma_{ff}} - \frac{1}{2}\right)^{-3}\right) \text{ for } 1.5 \leq \frac{z}{\sigma_{ff}} \leq \frac{L}{\sigma_{ff}} - 1.5 \quad (A5)$$

When $L/\sigma_{ff} \geq 3.0$, $a_{ads}(z)$ is given by,

$$\frac{a_{ads}(z)}{a(T)} = \frac{3}{8}\left(\frac{z}{\sigma_{ff}} + \frac{5}{6} - \frac{1}{3}\left(\frac{L-z}{\sigma_{ff}} - \frac{1}{2}\right)^{-3}\right) \text{ for } 0.5 \leq \frac{z}{\sigma_{ff}} \leq 1.5 \qquad (A6)$$

$$\frac{a_{ads}(z)}{a(T)} = \frac{3}{8}\left(\frac{8}{3} - \frac{1}{3}\left(\frac{z}{\sigma_{ff}} - \frac{1}{2}\right)^{-3} - \frac{1}{3}\left(\frac{L-z}{\sigma_{ff}} - \frac{1}{2}\right)^{-3}\right) \text{ for } 0.5 \leq \frac{z}{\sigma_{ff}} \leq \frac{L}{\sigma_{ff}} - 1.5$$

$$\qquad (A7)$$

$$\frac{a_{ads}(z)}{a(T)} = \frac{3}{8}\left(\frac{L-z}{\sigma_{ff}} + \frac{5}{6} - \frac{1}{3}\left(\frac{z}{\sigma_{ff}} - \frac{1}{2}\right)^{-3}\right) \text{ for } \frac{L}{\sigma_{ff}} - 1.5 \leq \frac{z}{\sigma_{ff}} \leq \frac{L}{\sigma_{ff}} - 0.5$$

$$\qquad (A8)$$

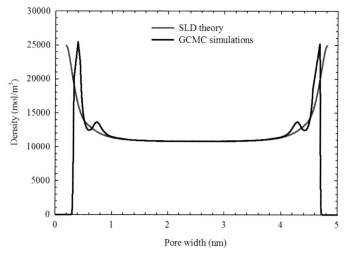

Fig. A1 Density profiles of CH_4 molecules in the 5-nm pore at 343.15 K and 30.0 MPa calculated from SLD theory and GCMC simulations.

Appendix 2: Density profile comparison between the SLD model and GCMC simulations

In Fig. A1, we calculate the density profiles of CH_4 molecules in the 5-nm pore at 343.15 K and 30.0 MPa using SLD theory and GCMC simulations. It is observed that density of CH_4 in the adsorption layer as well as at the pore center has a good agreement between the SLD model and the GCMC simulations, which can validate our SLD model in describing the distribution of CH_4 in nanopores.

References

Akkurt, R., Vinegar, H.J., Tutunjian, P.N., et al., 1996. NMR logging of natural gas reservoirs. Log. Anal. 37 (6), 33–42.

Alexeev, A.D., Ulyanova, E.V., Starikov, G.P., et al., 2004. Latent CH_4 in fossil coals. Fuel 83 (10), 1407–1411.

Ambrose, R.J., Hartman, R.C., Campos, M.D., et al., 2010. New pore-scale considerations for shale gas in place calculations. In: Presented at the SPE Unconventional Gas Conference, Pittsburgh, Pennsylvania, USA, 23–25 February.

Ambrose, R.J., Hartman, R.C., Diaz-Campos, M., Akkutlu, I.Y., Sondergeld, C.H., 2012. Shale gas-in-place calculations part I: new pore-scale considerations. SPE J. 17 (1), 219–229.

Anatoly, L., 2013. The basics of NMR. Magnetic Resonance Imaging for Groundwater. John Wiley & Sons, Inc.

Andersen, H.C., 1980. Molecular dynamics at constant pressure and/or temperature. J. Chem. Phys. 72, 2384.

Bhowmik, S., Dutta, P., 2011. Investigation into the methane displacement behavior by cyclic, pure carbon dioxide injection in dry, powdered, bituminous indian coals. Energy Fuel 25 (6), 2730.

Bloembergen, N., Purcell, E.M., Pound, R.V., 1948. Relaxation effects in nuclear magnetic resonance absorption. Phys. Rev. 73 (7), 679–712.

Brochard, L., Vandamme, M., Pellenq, R., et al., 2012. Adsorption-induced deformation of microporous materials: coal swelling induced by CO_2-CH_4 competitive adsorption. Langmuir 28, 2659.

Brunauer, S., Emmett, P.H., Teller, E., 1938. Adsorption of gases in multimolecular layers. J. Am. Chem. Soc. 60 (2), 309–319.

Bryan, W.P., 1987. Sorption hysteresis and the laws of thermodynamics. J. Chem. Educ. 64 (3), 209–212.

Busch, A., Gensterblum, Y., Krooss, B.M., 2003. Methane and CO_2 sorption and desorption measurements on dry argonne premium coals: pure components and mixtures. Int. J. Coal Geol. 55 (2), 205.

Cessford, N., Seaton, N., Düren T., 2012. Evaluation of ideal adsorbed solution theory as a tool for the design of metal-organic framework materials. Ind. Eng. Chem. Res. 51, 4911.

Chatterjee, R., Paul, S., 2013. Classification of coal seams for coal bed methane exploitation in central part of Jharia coalfield, India statistical approach. Fuel 111, 20.

Chen, J.H., Wong, D.S.H., Tan, C.S., Subramanian, R., Lira, C.T., Orth, M., 1997. Adsorption and desorption of carbon dioxide onto and from activated carbon at high pressures. Ind. Eng. Chem. Res. 36 (7), 2808–2815.

Chen, G., Zhang, J., Lu, S., Pervukhina, M., Liu, K., Xue, Q., 2016. Adsorption behavior of hydrocarbon on Illite. Energy Fuel 30 (11), 9114–9121.

Clarkson, C.R., Haghshenas, B., 2013. Modeling of supercritical fluid adsorption on organic rich shales and coal. In: SPE Unconv. Resour. Conf. Woodlands, Texas, USA.

Coates, G.R., Xiao, L.Z., Prammer, M.C., 1999. NMR Logging Principles and Applications. Gulf Publishing Company, Houston, TX.

Cristancho-Albarracin, D., Akkutlu, I.Y., Criscenti, L.J., Wang, Y., 2017. Shale gas storage in kerogen nanopores with surface heterogeneities. Appl. Geochem. 84, 1–10.

Dasani, D., Wang, Y., Tsotsis, T.T., Jessen, K., 2017. Laboratory-scale investigation of sorption kinetics of methane/ethane mixtures in shale. Ind. Eng. Chem. Res. 56, 9953–9963.

de Boer, J.H., Lippens, B.C., 1964. Studies on pore systems in catalysts II. The shapes of pores in aluminum oxide systems. J. Catal. 3 (1), 38–43.

Didar, B.R., Akkutlu, I.Y., 2013. Pore-size dependence of fluid phase behavior and properties in organic-rich shale reservoirs. In: SPE International Symposium on Oilfield Chemistry. Woodlands, Texas, USA.

Do, D.D., Do, H.D., 2003. Adsorption of supercritical fluids in non-porous and porous carbons: analysis of adsorbed phase volume and density. Carbon 41, 1777–1791.

Dobrzanski, C.D., Maximov, M.A., Gor, G.Y., 2018. Effect of pore geometry on the compressibility of a confined simple fluid. J. Chem. Phys. 148, 054503.

Dong, X., Liu, H., Hou, J., Wu, K., Chen, Z., 2016. Phase equilibria of confined fluids in nanopores of tight and shale rocks considering the effect of capillary pressure and adsorption film. Ind. Eng. Chem. Res. 55, 798–811.

Duan, S., Gu, M., Du, X., et al., 2016. Adsorption equilibrium of CO_2 and their mixture on Sichuan basin shale. Energy Fuel 30, 2248–2256.

Dubinin, M.M., 1960. The potential theory of adsorption of gases and vapors for adsorbents with energetically nonuniform surfaces. Chem. Rev. 60 (2), 235–241.

Errington, J.R., Panagiotopoulos, A.Z., 1999. A new intermolecular potential model for the n-alkane homologous series. J. Phys. Chem. B 103 (30), 6314–6632.

Faiz, M.M., Saghafi, A., Barclay, S.A., et al., 2007. Evaluating geological sequestration of CO_2 in bituminous coals: the Southern Sydney basin, Australia as a natural analogue. Int. J. Greenhouse Gas Control 1 (2), 223–235.

Fan, C., Do, D.D., Nicholson, D., 2011. On the cavitation and pore blocking in slit-shaped ink-bottle pores. Langmuir 27 (7), 3511–3526.

Firoozabadi, A., 2016. Thermodynamics and Applications in Hydrocarbon Energy Production. McGraw Hill, New York.

Fitzgerald, J.E., 1999. Adsorption of Pure and Multi-Component Gases of Importance to Enhanced Coaled Methane Recovery: Measurements and Simplified Local Density Model (Ph.D. dissertation). Oklahoma State University.

Fitzgerald, J.E., 2005. Adsorption of Pure and Multi-Component Gases of Importance to Enhanced Coalbed Methane Recovery: Measurements and Simplified Local-Density Modeling (Ph.D. dissertation). Oklahoma State University, Stillwater, OK.

Gasem, K.A.M., Gao, W., Pan, Z., Robinson, R.L., 2001. A modified temperature for the Peng-Robinson equation of state. Fluid Phase Equilib. 181 (1), 113–125.

Gasparik, M., Ghanizadeh, A., Bertier, P., Gensterblum, Y., Bouw, S., Krooss, B.M., 2012. High-pressure methane sorption isotherms of black shales from the Netherlands. Energy Fuel 26 (8), 4995–5004.

Gasparik, M., Bertier, P., Gensterblum, Y., et al., 2014. Geological controls on the methane storage capacity in organic-rich shales. Int. J. Coal Geol. 123, 34–51.

Gensterblum, Y., van Hemert, P., Billemont, P., Busch, A., Charriére, D., Li, D., 2009. European inter-laboratory comparison of high pressure CO_2 sorption isotherms. I: activated carbon. Carbon 47 (13), 2958–2969.

Gensterblum, Y., Busch, A., Krooss, B.M., 2014. Molecular concept and experimental evidence of competitive adsorption of H_2O, CO_2 and CH_4 on organic material. Fuel 115, 581.

Gerritsma, C.J., Osting, P.H., Trappeniers, N.J., 1971. Proton spin-lattice relaxation and self-diffusion in methanes: II experimental results for proton spin-relaxation times. Physica 51, 381–394.

Godec, M., Koperna, G., Petrusak, R., et al., 2013. Potential for enhanced gas recovery and CO_2 storage in the Marcellus Shale in the Eastern United States. Int. J. Coal Geol. 118, 95.

Gregg, S.J., Sing, K.S.W., 1982. Adsorption, Surface Area and Porosity. Academic Press, New York.

Groen, J.C., Peffer, L.A.A., Pérez-Ramı́rez, J., 2003. Pore size determination in modified micro- and mesoporous materials. pitfalls and limitations in gas adsorption data analysis. Microporous Mesoporous Mater. 60, 1–17.

Guo, R., Mannhardt, D., Kantzas, A., 2007. Characterizing moisture and gas content of coal by low-field NMR. J. Can. Pet. Technol. 46 (10), 49–54.

Halsey, G.D., 1948. Physical adsorption on non-uniform surfaces. J. Chem. Phys. 16, 931–937.

Heller, R., Zoback, M., 2014. Adsorption of methane and carbon dioxide on gas shale and pure mineral samples. J. Unconv. Oil Gas Res. 8, 14–24.

Hensen, E.J.M., Tambach, T.J., Bliek, A., et al., 2001. Adsorption isotherms of water in Li-, Na-, and K-montmorillonite by molecular simulations. J. Chem. Phys. 115 (7), 3322–3329.

Huang, L., Ning, Z., Wang, Q., et al., 2018a. Effect of organic type and moisture on CO_2/CH_4 competitive adsorption in kerogen with implications for CO_2 sequestration and enhance CH_4 recovery. Appl. Energy 210, 28.

Huang, L., Ning, Z., Wang, Q., et al., 2018b. Molecular simulation of adsorption behaviors of methane, carbon dioxide and their mixtures on kerogen: effect of kerogen maturity and moisture content. Fuel 211, 159.

Huang, X., Li, T., Gao, H., Zhao, J., Wang, C., 2019. comparison of so_2 with co_2 for recovering shale resources using low-field nuclear magnetic resonance. Fuel 245, 563–569.

Jiang, X., 2011. A review of physical modelling and numerical simulation of long-term geological storage of CO_2. Appl. Energy 88 (11), 3557.

Jiang, W., Lin, M., 2018. Molecular dynamics investigation of conversion methods for excess adsorption amount of shale gas. J. Nat. Gas Sci. Eng. 49, 241–249.

Jiang, J., Shao, Y., Younis, R.M., 2014. Development of a multi-continuum multicomponent model for enhanced gas recovery and CO_2 storage in fractured shale gas reservoirs. In: Proceedings of the SPE Improved Oil Recovery Symposium. Society of Petroleum Engineers, Tulsa, Oklahoma, April 12-16.

Jin, Z., Firoozabadi, A., 2013. Methane and carbon dioxide adsorption in clay-like slit pores by monte carlo simulations. Fluid Phase Equilib. 360, 456.

Jin, Z., Firoozabadi, A., September 2015. Phase behavior and flow in shale nanopores from molecular simulations. In: Presented at the SPE Annual Technical Conference and Exhibition in Houston, Texas, USA, 28-30.

Jin, Z., Firoozabadi, A., 2016a. Thermodynamic modeling of phase behavior in shale media. SPE J. 21 (1), 190.

Jin, Z., Firoozabadi, A., 2016b. Phase behavior and flow in shale nanopores from molecular simulations. Fluid Phase Equilib. 430, 156–168.

Jin, L., Ma, Y., Jamili, A., 2013. Investigating the effect of pore proximity on phase behavior and fluid properties in shale formations. Society of Petroleum Engineers, New Orleans, Louisiana.

Jin, B., Nasrabadi, H., 2016. Phase behavior of multi-component hydrocarbon systems in nano-pores using gauge-GCMC molecular simulation. Fluid Phase Equilib. 425, 324.

Jones, J.E., 1924. On the determination of molecular field-oii from the equation of state of a gas. Proc. R. Soc. London, Ser. A 106, 463.

Karacan, C.Ö., Ruiz, F., Cotè, M., Phipps, S., 2011. Coal mine methane: a review of capture and utilization practices with benefits to mining safety and to greenhouse gas reduction. Int. J. Coal Geol. 86, 121.

Kazemi, M., Takbiri-Borujeni, A., 2016. Molecular dynamics study of carbon dioxide storage in carbon-based organic nanopores. In: Presented at the SPE Annual Technical Conference and Exhibition in Dubai, UAE, September 26–28.

Keffer, D., Davis, H.T., McCormick, A., 1996. The effect of nanopore shape on the structure and isotherms of adsorbed fluids. Adsorption 2, 9–21.

Khosrokhavar, R., Wolf, K.H., Bruining, H., 2014. Sorption of CH_4 and CO_2 on a carboniferous shale from Belgium using a manometric setup. Int. J. Coal Geol. 128, 153.

Kim, J., Kim, D., Lee, W., Lee, Y., Kim, H., 2017a. Impact of total carbon and specific area on the adsorption capacity in horn river shale. J. Pet. Sci. Eng. 149, 331–339.

Kim, T.H., Cho, J., Lee, K.S., 2017b. Evaluation of CO_2 injection in shale gas reservoirs with multi-component transport and geomechanical effects. Appl. Energy 190, 1195.

Klomkliang, N., Do, D.D., Nicholson, D., 2013. On the hysteresis loop and equilibrium transition in slit-shaped ink-bottle pores. Adsorption 19 (6), 1273–1290.

Kowalczyk, P., Gauden, P.A., Terzyk, A.P., et al., 2012. Displacement of methane by coadsorbedcarbon dioxide is facilitated in narrow carbon nanopores. J. Phys. Chem. C 116 (25), 13640.

Kurniawan, Y., Bhatia, S.K., Rudolph, V., 2006. Simulation of binary mixture adsorption of methane and CO_2 at supercritical conditions in carbons. AICHE J. 52 (3), 957.

Langmuir, I., 1916. The constitution and fundamental properties of solids and liquids. Part I. Solids. J. Am. Chem. Soc. 38 (11), 2221.

Lee, L.L., 1988. Molecular Thermodynamics of Non-Ideal Fluids. Butterworths, Stoneham, MA.

Lemmon, E.W., McLinden, M.O., Friend, D.G., 2009. Thermophysical properties of fluid systems. In: NIST Chemistry WebBook, NIST Standard Reference Database Number 69. National Institute of Standards and Technology, Gaithersburg MD.

Li, Z., Firoozabadi, A., 2009. Interfacial tension of nonassociating pure substances and binary mixtures by density functional theory combined with peng-robinson equation of state. J. Chem. Phys. 130 (15), 154108.

Li, M., Gu, A., Lu, X., et al., 2002. Determination of the adsorbate density from supercritical gas adsorption equilibrium data. Carbon 41, 579–625.

Li, W.Z., Liu, Z.Y., Chen, Y.L., et al., 2007. Molecular simulation of adsorption and separation of mixtures of short linear alkanes in pillared layered materials at ambient temperature. J. Colloid Interface Sci. 312 (2), 179.

Li, Z., Jin, Z., Firoozabadi, A., 2014. Phase behavior and adsorption of pure substances and mixtures and characterization in nanopore structures by density functional theory. SPE J. 19 (6), 1096.

Li, Y., Wang, S., Wang, Q., et al., 2016. Molecular dynamics simulations of tribology properties of NBR (nitrile-butadiene rubber)/carbon nanotube composites. Compos. Part B 97, 62.

Liu, Y., Wilcox, J., 2012. Molecular simulations of CO_2 adsorption in micro- and mesoporous carbons with surface heterogeneity. Int. J. Coal Geol. 104, 83.

Liu, F., Ellett, K., Xiao, Y., et al., 2013. Assessing the feasibility of CO_2 storage in the new Albany shale (Devonian-Mississippian) with potential enhanced gas recovery using reservoir simulation. Int. J. Greenhouse Gas Control 17, 111.

Liu, X., He, X., Qiu, N., et al., 2016. Molecular Simulation of CH_4, CO_2, H_2O and N_2 molecules adsorption on heterogeneous surface models of coal. Appl. Surf. Sci. 389, 894.

Liu, Y., Jin, Z., Li, H., 2017. Comparison of PR-EOS with capillary pressure model with engineering density functional theory on describing the phase behavior of confined hydrocarbons. In: Presented at the SPE Annual Technical Conference and Exhibition in San Antonio, Texas, USA, 9-11 October.

Liu, Y., Jin, Z., Li, H.A., 2018a. Comparison of Peng-Robinson equation of state with capillary pressure model with engineering density-functional theory in describing the phase behavior of confined hydrocarbons. SPE J. 23 (05), 1784.

Liu, Y., Li, H., Tian, Y., Jin, Z., Deng, H., 2018b. Determination of the absolute adsorption/desorption isotherms of CH_4 and n-C_4H_{10} on shale from a nano-scale perspective. Fuel 218, 67–77.

Lu, X.C., Li, F.C., Watson, A.T., 1995. Adsorption measurements in Devonian shales. Fuel 74 (4), 599–603.

Lu, L., Wang, Q., Liu, Y., 2003. Adsorption and separation of ternary and quaternary mixtures of short linear alkanes in zeolites by molecular simulation. Langmuir 19 (25), 10617.

Lu, X., Jin, D., Wei, S., et al., 2015. Competitive adsorption of a binary CO_2-CH_4 mixture in nanoporous carbons: effects of edge-functionalization. Nanoscale 7 (3), 1002.

Luo, F., Xu, R.N., Jiang, P.X., 2013. Numerical investigation of the influence of vertical permeability heterogeneity in stratified formation and of injection/production well perforation placement on CO_2 geological storage with enhanced CH_4 recovery. Appl. Energy 102, 1314.

Majewska, Z., Ceglarska-Stefańsk, G., Majewski, S., et al., 2009. Binary gas sorption/desorption experiments on a bituminous coal: simultaneous measurements on sorption kinetics, volumetric strain and acoustic emission. Int. J. Coal Geol. 77 (1), 90.

Martin, M.G., Siepmann, J.I., 1998. Transferable potentials for phase equilibria. 1. united-atom description of n-alkanes. J. Phys. Chem. B 102 (14), 2569–2577.

Megawati, M., Madland, M.V., Hiorth, A., 2012. Probing pore characteristics of deformed chalk by NMR relaxation. J. Pet. Sci. Eng. 100, 123–130.

Menon, P.G., 1968. Adsorption at high pressures. J. Phys. Chem. 72, 2695–2696.

Metropolis, N., Rosenbluth, A.W., Rosenbluth, M.N., et al., 1953. Equation of state calculations by fast computing machines. J. Chem. Phys. 21 (6), 1087–1092.

Mohammad, S.A., Chen, J.S., Robinson Jr., R.L., Gasem, K.A.M., 2009. Generalized simplified local-density/Peng-Robinson model for adsorption of pure and mixed gases on coals. Energy Fuel 23, 6259–6271.

Mohammad, S.A., Arumugam, A., Robinson Jr., R.L., Gasem, K.A.M., 2011. High-pressure adsorption of pure gases on coals and activated carbon: measurements and modeling. Energy Fuel 26 (1), 536–548.

Myers, A.L., Prausnitz, J.M., 1965. Thermodynamics of mixed-gas adsorption. AICHE J. 11 (1), 121.

Neimark, A.V., Vishnyakov, A., 2000. Gauge cell method for simulation studies of phase transitions in confined systems. Phys. Rev. E 62 (4), 4611.

Ottiger, S., Pini, R., Storti, G., et al., 2008. Measuring and modeling the competitive adsorption of CO_2, CH_4, and N_2 on a dry coal. Langmuir 24 (17), 9531.

Ozawa, S., Kusumi, S., Ogino, Y., 1976. Physical adsorption of gases at high Pressure. IV. An improvement of the Dubinin-Astakhov adsorption equation. J. Colloid Interface Sci. 56, 83–91.

Pang, Y., Mohamed, Y.S., Sheng, J., 2018. Investigating gas-adsorption, stree-dependense, and non-Darcy-flow effects on gas storage and transfer in nanopores by use of simplified local density model. SPE Reserv. Eval. Eng. 21 (1), 73–95.

Pedram, E.O., Hines, A.L., Cooney, D.O., 1984. Adsorption of light hydrocarbons on spent shale produced in a combustion retort. Chem. Eng. Commun. 27, 181–191.

Peng, D., Robinsion, D.B., 1976. A new two-constant equation of state. Ind. Eng. Chem. Fundam. 15 (1), 59–64.

Rangarajan, B., Lira, C.T., Subramanian, R., 1995. Simplified local density model for adsorption over large pressure ranges. AICHE J. 41 (4), 838–845.

Rao, Z., Wang, S., Wu, M., et al., 2012. Molecular dynamics simulations of melting behavior of alkane as phase change materials slurry. Energy Convers. Manag. 64 (4), 152–156.

Rexer, T.F.T., Benham, M.J., Aplin, A.C., Thomas, K.M., 2013. Methane adsorption on shale under simulated geological temperature and pressure conditions. Energy Fuel 27, 3099–3109.

Riewchotisakul, S., Akkutlu, I.Y., 2016. Adsorption-enhanced transport of hydrocarbons in organic nanopores. SPE J. 21 (06), 1960–1969.

Rigby, D., Sun, H., Eichinger, B.E., 1997. Computer simulations of poly (ethylene oxide): force field, PVT diagram and cyclization behaviour. Polym. Int. 44 (3), 311.

Roque-Malherbe, R.M.A., 2007. Adsorption and Diffusion in NanoporousMaterials. CRC Press, Boca Raton, FL.

Ross, D.J., Bustin, R.M., 2007. Impact of mass balance calculations on adsorption capacities in microporous shale gas reservoirs. Fuel 86 (17), 2696.

Ross, D.J.K., Bustin, R.M., 2009. The importance of shale composition and pore structure upon gas storage potential of shale gas reservoirs. Mar. Pet. Geol. 26, 916–927.

Santos, J.M., Akkutlu, I.Y., 2013. Laboratory measurement of sorption isotherm under confining stress with pore-volume effects. SPE J. 18 (5), 924–931.

Shirono, K., Daiguiji, H., 2007. Molecular simulation of the phase behavior of water confined in silica nanopores. J. Phys. Chem. C 111 (22), 7938.

Sing, K.S.W., Everett, D.H., Haul, R.A.W., et al., 2008. Reporting physisorption data for gas/solid systems. In: Handbook of Heterogeneous Catalysis. Wiley-VCH Verlag GmbH & Co. KGaA.

Singh, S.K., Sinha, A., Deo, G., et al., 2009. Vapor liquid phase coexistence, critical properties, and surface tension of confined alkanes. J. Phys. Chem. C 113 (17), 7170.

Smit, B., Karaborni, S., Siepmann, J.I., 1995. Computer simulations of vapor-liquid phase equilibria of n-alkanes. J. Chem. Phys. 102 (5), 2126–2140.

Song, Y., Jiang, B., Li, W., 2017. Molecular simulations of $CH_4/CO_2/H_2O$ competitive adsorption on low rank coal vitrinite. Phys. Chem. Chem. Phys. 19, 17773.

Steele, W.A., 1973. The physical interaction of gases with crystalline solids: I. Gas-solid energies and properties of isolated adsorbed atoms. Surf. Sci. 36 (1), 317–352.

Sun, H., 1998. Compass: an ab initio force-field optimized for condensed-phase applications overview with details on alkane and benzene compounds. J. Phys. Chem. B 102 (38), 7338–7373.

Sun, H., Zhao, H., Qi, N., et al., 2017. Molecular insights into the enhanced shale gas recovery by carbon dioxide in kerogen slit nanopores. J. Phys. Chem. C 121 (18), 10233.

Sweatman, M.B., Quirke, N., 2002. Predicting the adsorption of gas mixtures: adsorbed solution theory versus classical density functional theory. Langmuir 18 (26), 10443.

Ta, T.D., Tieu, A.K., Zhu, H., et al., 2015. Adsorption of normal-alkanes on Fe (110), FeO (110), and $Fe_2O_3(0001)$: influence of iron oxide surfaces. J. Phys. Chem. C 119 (23), 12999–13010.

Tarazona, P., Marconi, U., Evans, R., 1987. Phase equilibria of fluid interfaces and confined fluids. Mol. Phys. 60 (3), 573–595.

Tian, Y., Yan, C., Jin, Z., 2017. Characterization of methane excess and absolute adsorption in various clay nanopores from molecular simulation. Sci. Rep. 7, 12040.

Tsai, M.C., Chen, W.N., Cen, P.L., et al., 1985. Adsorption of gas mixture on activated carbon. Carbon 23 (2), 167–173.

Valentini, P., Schwartzentruber, T.E., Cozmuta, I., 2011. ReaxFFGrand canonical Monte Carlo simulation of adsorption and dissociation of oxygen on platinum (111). Surf. Sci. 605 (23), 1941.

van der Ploeg, P., Berendsen, H.J.C., 1982. Molecular dynamics simulations of a bilayer membrane. J. Chem. Phys. 76 (6), 3271–3276.

van der Sommen, J., Zwietering, P., Eillebrecht, B.J.M., et al., 1955. Chemical structure and properties of coal X2-adsorption capacity for CH_4. Fuel 34, 444–448.

Vishal, V., Singh, T.N., Ranjith, P.G., 2015. Influence of sorption time in CO_2-ECBM process in Indian coals using coupled numerical simulation. Fuel 139, 51.

Walton, J.P.R.B., Quirke, N., 1989. Capillary condensation: a molecular simulation study. Mol. Simul. 2 (4–6), 361.

Wang, Y., Tsotsis, T.T., Jessen, K., 2015. Competitive sorption of methane/ethane mixtures on shale: measurements and modeling. Ind. Eng. Chem. Res. 54, 12178.

Wang, X., Zhai, Z., Jin, X., et al., 2016a. Molecular simulation of CO_2/CH_4 competitive adsorption in organic matter pores in shale under certain geological conditions. Pet. Explor. Dev. 43 (5), 841.

Wang, Y., Zhu, Y., Liu, S., et al., 2016b. Methane adsorption measurements and modeling for organic-rich marine shale samples. Fuel 172, 301–309.

Weijermars, R., 2013. Economic appraisal of shale gas plays in continental Europe. Appl. Energy 106, 100.

Weijermars, R., 2014. US shale gas production outlook based on well roll-out rate scenarios. Appl. Energy 124, 283.

Weniger, P., Kalkreuth, W., Busch, A., et al., 2010. High-pressure methane and carbon dioxide sorption on coal and shale samples from the Parana Basin, Brazil. Int. J. Coal Geol. 84 (3–4), 190.

White, C.M., Smith, D., Jones, K., et al., 2005. Sequestration of carbon dioxide in coal with enhanced coalbed methane recovery-a review. Energy Fuel 19 (3), 659.

Widom, B., 1963. Some topics in the theory of fluids. J. Chem. Phys. 39 (11), 2808–2812.

Wongkoblap, A., Do, D.D., Birkett, G., et al., 2011. A critical assessment of capillary condensation and evaporation equations: a computer simulation study. J. Colloid Interface Sci. 356 (2), 672.

Wu, H., Chen, J., Liu, H., 2015a. Molecular dynamics simulations about adsorption and displacement of methane in carbon nanochannels. J. Phys. Chem. C 2119 (24), 13652.

Wu, Y., Fan, T., Jiang, S., et al., 2015b. Methane adsorption capacities of the lower paleozoic marine shales in the Yangtze Platform, South China. Energy Fuel 29, 4160–4167.

Xiong, F., Wang, X., Amooie, N., et al., 2017. The shale gas sorption capacity of transitional shales in the Ordos Basin, NW China. Fuel 208, 236–246.

Yamazaki, T., Aso, K., Chinju, J., 2006. Japanese potential of CO₂ sequestration in coal seams. Appl. Energy 83 (9), 911.

Yao, Y.B., Liu, D.M., et al., 2010. Petrophysical characterization of coals by low-field nuclear magnetic resonance (NMR). Fuel 89, 1371–1380.

Yao, Y., Liu, D., Xie, S., 2014. Quantitative characterization of CH₄ adsorption on coal using a low-field NMR relaxation method. Int. J. Coal Geol. 131, 32–40.

Yaws, C.L., 2003. Yaws' Handbook of Thermodynamic and Physical Properties of Chemical Compounds. Knovel Corporation, New York, NY.

Yu, W., Al-Shalabi, E.W., Sepehrnoori, K., 2014. A sensitivity study of potential CO₂ injection for enhanced gas recovery in Barnett shale reservoirs. In: Proceedings of the SPE Unconventional Resources Conference. Society of Petroleum Engineers, Woodlands, Texas, April 1–3.

Yu, S., Yanming, Z., Wu, L., 2017. Macromolecule simulation and CH₄ adsorption mechanism of coal vitrinite. Appl. Surf. Sci. 396, 291.

Yuan, J., Luo, D., Feng, L., 2015a. A review of the technical and economic evaluation techniques for shale gas development. Appl. Energy 148, 49.

Yuan, Q., Zhu, X., Lin, K., et al., 2015b. Molecular dynamics simulations of the enhanced recovery of confined methane with carbon dioxide. Phys. Chem. Chem. Phys. 17, 31887.

Zhang, J., Liu, K., Clennell, M.B., et al., 2015. Molecular simulation of CO₂-CH₄ competitive adsorption and induced coal swelling. Fuel 160, 309.

Zhao, Y., Feng, Y., Zhang, X., 2016. Molecular simulation of CO₂/CH₄ self- and transport diffusion coefficients in coal. Fuel 165, 19.

Zhou, S., Wang, H., Xue, H., Guo, W., Lu, B., 2016. Difference between excess and absolute adsorption capacity of shale and a new shale gas reserve calculation method. Nat. Gas Ind. 36, 12–20.

CHAPTER FOUR

Interfacial tension for CO_2/CH_4/ brine systems under reservoir conditions

Shale gas is playing an increasingly important role in global energy portfolio since 2010; it accounted for 23% of total world energy supply in 2010 and will reach 49% in 2035, according to the report on annual outlook of global energy from US energy information administration (EIA). Recent years have witnessed an increasing trend in developing new technologies for recovering the vast shale gas resources around the globe, such as hydraulic fracturing techniques. Water-less fracturing, such as CO_2 fracturing, has attracted an extensive attention because of the unique properties of CO_2, such as a higher Langmuir adsorption in shale matrix compared to CH_4 (Heller and Zoback, 2014), the compatibility between CO_2 and reservoir fluids (CH_4 and water), and large diffusivity of CO_2 in shale pores. These properties might enable CO_2-based fracturing technique to mitigate the for-mation damage issues that are otherwise caused by water-based fracturing, hence promoting a higher gas recovery post fracturing. Enhancing shale gas recovery through injecting CO_2 is also under investigation in some shale reservoirs (Hussen et al., 2012). Additional benefits of using CO_2 include storing CO_2 in shale formations. Either CO_2-based fracturing or CO_2-based enhanced gas recovery requires a profound understanding on the phase behavior and interfacial properties of the CO_2/CH_4/brine systems under reservoir conditions (Li and Elsworth, 2014).

Interfacial tension (IFT) of gas-water or gas-brine is one of the most important properties affecting the performance of enhanced gas recovery. It significantly affects the movement, phase behavior, and distribution of res-ervoir fluids in porous media (Danesh, 1998). Specifically, optimum oper-ations of CO_2 flooding and sequestration in oil/gas reservoirs also depend on accurate knowledge of IFT of CO_2/brine systems, which affects the trans-port properties and capillary-sealing efficiency of CO_2 in the formation (Li et al., 2013; Chalbaud et al., 2006; Aggelopoulos et al., 2010; Shah et al., 2008).

Confined Fluid Phase Behavior and CO2
Sequestration in Shale Reservoirs
https://doi.org/10.1016/B978-0-323-91660-8.00008-7

There have been extensive experimental and modeling studies on quantifying the IFT of various gas–water systems. Axisymmetric drop shape analysis method (ADSA) is the most-widely used technique to perform IFT measurement. With the ADSA method, IFT is measured by solving the Young-Laplace equation based on the geometry of a pendant drop captured by the measurement (Bahramian and Danesh, 2004, 2005). Table 4.1 summarizes some of the relevant gas–water IFT measurements and the range of laboratory conditions under which the measurements were conducted. It can be seen from Table 4.1 that extensive experimental studies have been conducted on pure gas-pure water systems over wide ranges of pressures

Table 4.1 Summary of previous laboratory measurements on gas-water IFT.

References	System compositions	Temperature range (°F)	Pressure range (psia)
Hocott (1939)	$CH_4/C_2H_6/C_3H_8/H_2O$	78.0–150.0	14.5–3510
Hough et al. (1951)	CH_4/H_2O	74.0–280.0	15–15,000
Heuer (1957)	CO_2/H_2O	100.0–280.0	Up to 10,000
Jennings and Newman (1971)	CH_4/H_2O	74.0, 212.0, 350.0	14.7–12,000
Massoudi and King Jr. (1974)	CH_4/H_2O, CO_2/H_2O, N_2/H_2O	77.0	up to 1000
Jho et al. (1978)	CO_2/H_2O	32.0–122.0	60–1000
Wiegand and Franck (1994)	$CH_4/C_3H_8/C_6H_{14}/C_{10}H_{22}/N_2/H_2O$, etc.	77.0–571.0	14.5–37,710
Chun and Wilkinson (1995)	CO_2/H_2O/ethanol	41.0–160.0	14.5–2700
Sachs and Meyn (1995)	CH_4/H_2O	77.0	58–6802
Lepski (1997)	CH_4/H_2O, N_2/H_2O	126.5–260.2	1500–3500
Tian et al. (1997)	CH_4/H_2O, C_6H_{14}/H_2O, C_7H_{16}/H_2O, N_2/H_2O, etc.	76.7–400.0	14.7–29,008
da Rocha et al. (1999)	CO_2/H_2O	95.0–149.0	1000–4000
Ren et al. (2000)	CH_4/H_2O, $CH_4/CO_2/H_2O$	77.0–212.0	145–4351
Yan et al. (2001)	$CH_4/N_2/H_2O$, $CO_2/N_2/H_2O$	77.0–212.0	145–4351
Hebach et al. (2002)	CO_2/H_2O	41.0–144.0	14.5–2900
Zhao (2002)	CH_4/H_2O	77.0–212.0	145–4351
Tewes and Boury (2005)	CO_2/H_2O	68.0, 86.0, 104.0	290–1305

Table 4.1 Summary of previous laboratory measurements on gas-water IFT—cont'd

References	System compositions	Temperature range (°F)	Pressure range (psia)
Park et al. (2005)	CO_2/H_2O	68.0, 77.0, 100.4, 159.8	Up to 2941
Yang et al. (2005)	CO_2/brine	77.0–136.0	14.5–4351
Chiquet et al. (2007)	CO_2/H_2O	95.0–230.0	725–6527
Akutsu et al., 2007	CO_2/H_2O	77.0, 95.0, 113.0	1088–2393
Sutjiadi-Sia et al. (2008)	CO_2/H_2O	104.0	Up to 3916
Bennion and Bachu (2008)	CO_2/H_2O/brine	105.0–257.0	290–3916
Rushing et al. (2008)	CH_4/C_2H_6/C_3H_8/N_2/CO_2/ H_2O	300.0–400.0	1000–20,000
Bachu and Bennion (2009)	CO_2/H_2O/brine	68.0–257.0	290–3916
Aggelopoulos et al. (2010)	CO_2/brine	81.0–212.0	725–3626
Georgiadis et al. (2010)	CO_2/H_2O	77.0–214.0	145–4351
Chalbaud et al. (2010)	CO_2/brine	81.0–212.0	3771
Shariat et al. (2011)	CH_4/C_2H_6/C_3H_8/H_2O	300.0–400.0	1000–20,000
Aggelopoulos et al. (2011)	CO_2/brine	80.6, 159.8, 212.0	725–3626
Shariat et al. (2012)	CO_2/H_2O	Up to 400.0	1000–18,000
Li et al. (2012a,b)	CO_2/brine	76.7–346.7	290–7252
Li et al. (2012a,b)	CO_2/brine	157.7–301.7	290–7252
Khosharay and Varaminian (2014)	CH_4/H_2O, C_2H_6/H_2O, CO_2/H_2O,C_3H_8/H_2O	51.8–102.2	Up to 870
Pereira et al. (2016)	CO_2/H_2O	76.7–384.5	49–10,028
Khashefi et al. (2016)	CH_4/H_2O, CH_4/brine	100.1–391.7	0–13,343

and temperatures. Most of the existing studies did not address the effects of nonhydrocarbon contaminants on gas-water IFT, especially at high-pressure/temperature reservoir conditions. Moreover, most of the gas/water IFT measurements are only made for the pure hydrocarbon gases, rather than gas mixtures, with water or brine. Ren et al. (2000) measured the interfacial tension of $CH_4/CO_2/H_2O$ systems. They covered the temperature range of 76.7–211.7°F and pressure range of 145–4351 psia. But the salinity effect on the IFT was not addressed.

In shale formations, the presence of salinity can affect the IFT of reservoir fluids to a large extent. It has been recognized that the addition of salts into the aqueous phase can significantly increase the IFT of gas/brine systems (Massoudi and King, 1975; Li et al., 2012a,b). Some of the previous studies attributed the salinity effect to the change of the interface structure: The cations tend to accumulate in the aqueous phase due to the adsorption of the cations on the interface (Khashefi et al., 2016; Ralston and Healy, 1973; Johansson and Eriksson, 1974; Pegram and Record, 2008; Levin et al., 2009). Another reason causing the IFT increase is the density increase of the aqueous phase because of salt addition. Yang et al. (2005) reported IFT for CO_2/brine system over 77.0–136.0°F and 14.5–4351 psia. Bennion and Bachu (2008) measured the IFT for CO_2/brine system over 105.0–257.0°F and 290–3916 psia. Aggelopoulos et al. (2010) presented the IFT data of CO_2/brine system, with the consideration of different concentrations of NaCl and $CaCl_2$. Chalbaud et al. (2010) measured the IFT for CO_2/brine systems at salinities of 0.085–2.75 mol/kg. Khashefi et al. (2016) carried out IFT measurements on CH_4/brine and CH_4/pure water systems using the ADSA method in the temperature range of 100.1–391.7°F and at pressures up to 13,343 psia. Bachu and Bennion (2009) conducted the IFT measurement of CO_2/water and CO_2/brine systems over 68.0–257.0°F and 290–3916 psia. Li et al. (2012a,b) measured the IFT between CO_2 with different salts in a wide range of total salt molality. Nonetheless, the experimental data for IFT of CH_4/brine mixtures are limited. Meanwhile, experimental data for IFT of CO_2/CH_4/brine mixtures are still scarce at reservoir conditions, albeit extensive IFT measurements have been conducted for CO_2/brine mixtures in the past decades.

An accurate IFT model is needed to predict the IFT of gas/brine systems under reservoir conditions. Up to now, numerous correlations were proposed and some of them have been used in commercial reservoir simulators for estimating IFT by petroleum engineering industry. The Parachor model (Weinaug and Katz, 1943; Macleod, 1923) and the scaling law (Lee and Chien, 1984) have gained more use than other predictive methods because of their simplicity (Danesh, 1998). However, both methods are not recommended for the IFT predictions of hydrocarbon/water systems (Danesh, 1998). Massoudi and King Jr. (1974) presented an IFT correlation for pure CO_2/water systems considering pressure and temperature; but it can be only applied at one temperature. Firoozabadi and Ramey Jr. (1988) proposed an IFT model that can predict the IFT of hydrocarbon-gas/water mixtures. Argaud (1992) and Sutton (2009) developed new IFT correlations based on the Firoozabadi and Ramey Jr. (1988) model by considering a broader class of compounds. Argaud (1992) added the ratio of Parachor to molar mass of each compound to the Firoozabadi and Ramey Jr. (1988) correlation as a corrective factor, while Sutton (2009) considered more parameters in the improved correlation. Nonetheless, the predictive capabilities of these improved models are still limited (Johansson and Eriksson, 1974). Bennion and Bachu (2008) presented an IFT correlation between CO_2 and brine as a function of salinity, which predicts the IFT of CO_2/brine systems based on the solubility of CO_2 in brine. However, the correlation of Bennion and Bachu (2008) cannot predict IFT at pressures and temperatures higher than 3916 psia and 257.0°F. Meanwhile, the correlation was developed based on their own measured data, without being validated by other experimental data. Hebach et al. (2002) and Kvamme et al. (2007) presented IFT correlations for CO_2/water mixtures considering reservoir temperature, pressure, and density differences of pure component, but excluding the effect of mutual solubility. Furthermore, Li et al. (2012a,b) and Chalbaud et al. (2009) developed correlations for IFT of CO_2/brine mixtures. Other methods based on statistical thermodynamics were also applied to predict IFT, such as linear gradient theory (Yan et al., 2001), perturbation theory (Nordholm et al., 1980), density gradient theory (DGT) (Cahn and Hilliard, 1958; Rowlinson, 1979), and integral and density functional theories (Evans, 1979; Almeida and Telo da Gama, 1989; Bongiorno and Davis, 1975). In general, these methods have not been widely used in the petroleum industry likely due to their complexity.

In this section, previous IFT measurements of the gas/water or gas/brine mixtures are first reviewed and summarized. New experimental IFT data for CO_2/CH_4/brine systems with NaCl concentrations up to 200,000 ppm are presented over 77.0–257.0°F and 15–5027 psia. IFT data for CH_4/water and CO_2/water mixtures are found to be in good agreement with published data. The effects of temperature, pressure, CO_2 concentration, and salinity on IFT of CO_2/CH_4/brine mixtures are examined in detail. Based on the measured IFT data, a new IFT model is developed to determine IFT of CO_2/CH_4/brine mixtures. This new model's performance is examined by comparing it with other commonly used IFT correlations.

4.1 ADSA IFT apparatus

Fig. 4.1 shows an image of the experimental setup used for the ADSA IFT measurements. The major component of this system is a visual high-pressure cell (TEMCO, Inc., United States) with a chamber volume of approximately $41.5\,cm^3$. It can sustain pressure up to 10,130.9 psia and temperature up to 350.0°F. A light source was used to illuminate the pendant drop in the glass-windowed chamber. Nitrile O-rings were used in this experiment to reduce the corrosion of O-rings caused by CO_2 exposure. A band heater, together with an insulation jacket and a resistance temperature device (RTD) sensor, was used to heat the IFT cell and control its temperature within $\pm0.1\,K$. The IFT cell was placed on a vibration-free Table (RS4000, United States) to remove the effect of constant low-frequency vibration. A needle valve was employed for controlling the formation of pendant drop, while several other valves were used to control the introduction of the different fluids (e.g., CO_2 or CH_4) into the pressure cell. The drain valve and a needle cleanout valve were used to flush and clean the cell chamber and needle without removal of the glass windows. A high-resolution camera was used to observe the formation of the pendant drop, and capture its image. The stainless-steel needles could be changed to cover different IFT measurement ranges.

Fig. 4.2 shows the schematic of the ADSA experimental setup used in this study. The pressure of the high-pressure IFT cell was measured with a digital precision testing gauge (DPG409-5.0kG, Ashcroft) with an accuracy of 0.05% of the full range. The temperature was measured with a

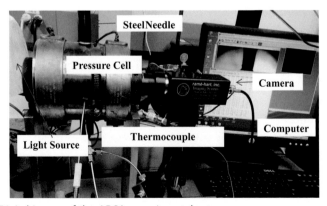

Fig. 4.1 Digital image of the ADSA experimental setup.

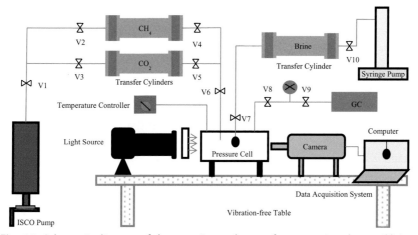

Fig. 4.2 Schematic diagram of the experimental setup for measuring the equilibrium IFTs of CO$_2$/CH$_4$/brine systems using the ADSA technique for the pendant drop case.

thermocouple (JMQSS-125U-6, Omega) with an accuracy of ± 0.1 K. The LED light source with a glass diffuser was used to provide a uniform illumination for the pendant drop. Two transfer cylinders, connected to the IFT cell, were used to pressurize and inject CH$_4$ and CO$_2$. The pressure of transfer cylinders was controlled by a syringe pump (500 HP, ISCO, Inc., Lincoln, NE). The pressure measurement, temperature measurement, and determination of mixture composition have accuracies of ± 3 psia, ± 0.1 K, and ± 3.0 wt%, respectively. Considering the inaccuracies that arise from the ASDA method as well as from the estimated phase densities, a conservative uncertainty of $\pm 5\%$ can be applied to the experimentally measured IFTs. The IFT of the CO$_2$/CH$_4$/brine systems is measured over 77.0–257.0°F, 15–5027 psia, and a salinity range of 0–200,000 ppm of NaCl. Each IFT measurement was repeated three times to ensure the repeatability of each measurement.

Before each measurement, the entire system was tested for leakage with N$_2$. Then it was cleaned with acetone, flushed with CH$_4$ or CO$_2$ and evacuated. The cell was pressurized with CH$_4$ or CO$_2$ to a prespecified pressure. When measuring the IFT for gas mixtures, the pressure cell was first filled with a pure gas (e.g., CO$_2$) to a specified pressure at a given temperature; then another pure gas (e.g., CH$_4$) was injected into the pressure cell, resulting in a different pressure. A sampler (Swagelok, Canada) with a volume of 10 cm^3 was used to take the gas sample inside the pressure cell. The composition of the gas mixture was measured with a gas chromatography

(GC) method. After the pressure and temperature in the pressure cell were stabilized, a pendant water/brine drop was introduced by a syringe pump (500 HP, ISCO, Inc., Lincoln, NE), of which pressure was maintained about 14–44 psia higher than that of gas phase inside the pressure cell. The pendant water drop formed at the tip of the stainless-steel needle. After the gas was injected, usually 30–60 min were required for the system to reach an equilibrium state at given pressure and temperature.

After the pendant water drop was formed in the gas phase, its digital image was well-focused through the diffused light, acquired sequentially, and stored by the computer. For each digital water drop image, a standard grid image was used to calibrate the drop image and correct possible optical distortion. The ADSA program for the pendant drop case was then executed to determine the equilibrium IFT. The output data also included the radius of the curvature at the apex point, and the volume and surface area of the pendant water drop. Only the local gravitational acceleration and the gas-water density difference were required as the input data for this program. Knowing the pendant drop dimensions and the fluid densities enabled the calculations of IFT. During the IFT measurement, gas-phase and liquid-phase densities needed to be input into the software. In this study, as for CO_2/H_2O, CH_4/H_2O, and $CO_2/CH_4/H_2O$ systems, we calculated the densities of the liquid phase and vapor phase by an enhanced Peng-Robinson equation of state (PR EOS) model with temperature-dependent binary interaction parameters and constant volume shift parameters. More specifically, a new BIP correlation is used developed by Li and Yang (2013) to estimate the BIP of CO_2/H_2O binary; this BIP correlation is a function of the reduced temperature of CO_2. Meanwhile, we used a BIP correlation developed by Søreide and Whitson (1992) to estimate the BIP of CH_4/H_2O binary; this BIP correlation is a function of temperature and acentric factor of CH_4. Table 4.2 lists the physical properties of CO_2, CH_4 and H_2O used in the PR EOS model. As for CO_2/brine, CH_4/brine, and CO_2/CH_4/brine systems, in order to obtain an accurate phase-density predictions, we used a modified PR EOS model by Søreide and Whitson

Table 4.2 Physical properties of the three components used in the IFT model.

Component	P_c (psia)	T_c (°F)	Acentric factor	Molecular weight	Volume shift	Parachor
CO_2	1069.9	87.89	0.225	44.01	−0.15400	78
CH_4	667.2	−116.59	0.008	16.04	−0.01478	77
H_2O	3197.8	705.47	0.344	18.02	0.23170	52

(1992) with constant volume shift parameters. This model considers salinity and mutual solubility of CH_4/brine and CO_2/brine binaries.

In this section, much care has been taken to eliminate possible error sources in IFT measurements. Firstly, as recommended by the manufacturer, the settings for KRÜSS software suitable for gas-water IFT measurements were set as (Light level = 2, Brightness = 31, Gain = 10, Exposure = − 11). Secondly, a steel needle with an outer needle diameter of 0.70 mm was used in the tests to control the droplet size. During the experiments, extra efforts were devoted to generating large droplets at the needle tip; larger droplet volumes created more accurate IFT measurements because the effect of the capillary tube tip diminished as the drop volume became larger. In addition, all the IFT data were measured under equilibrium conditions.

4.2 Mathematical formulation

Most of the previous IFT models originated from the Parachor model (Chalbaud et al., 2006; Li et al., 2012a,b; Zuo and Stenby, 1997; Sato, 2003; Ayirala and Rao, 2004; Fawcett, 1994; Hough and Stegemeier, 1961; Ayirala and Rao, 2006). For example, Chalbaud et al. (2006) developed a correlation on the basis of the Parachor model taking into account the influence of temperature, pressure, salt presence and chemical structure of CO_2. Ayirala and Rao (2006) proposed a new mechanistic Parachor model based on mass transfer to predict IFT in multicomponent hydrocarbon systems.

Sudgen (1921) proposed an equation including the new constant Parachor in the following form:

$$\sigma^{1/4} = \frac{P}{M}\Delta\rho \tag{4.1}$$

where σ is the IFT between two phases; P is Parachor; M is molecular weight of the component; and $\Delta\rho$ is density difference between two phases. Quayle (1953) determined the Parachor for a large number of compounds considering their molecular structures. Weinaug and Katz (1943) extended Sudgen's equation (Sudgen, 1921) to mixtures as follows:

$$\sigma^{1/4} = \sum_{i=1}^{n} P_i \left(\frac{\rho_L}{M_L} x_i - \frac{\rho_v}{M_v} y_i \right) \tag{4.2}$$

where P_i is Parachor for component i; M_L is the average molecular weight of liquid phase; M_V is the average molecular weight of vapor phase; ρ_L is density of liquid phase; ρ_V is density of vapor phase; x_i is the mole fraction of component i in liquid phase; and y_i is the mole fraction of component i in vapor phase. The equation proposed by Weinaug and Katz (1943) is used as a standard method of IFT prediction in the petroleum industry. It has been applied to some binary hydrocarbon systems and pure hydrocarbons successfully, but generally does not perform well for gas/water systems (Firoozabadi and Ramey Jr., 1988).

Firoozabadi and Ramey Jr. (1988) presented a correlation for estimating the IFT of hydrocarbon gas or hydrocarbon liquid with water. The phase-density difference and reduced temperature for the hydrocarbon phase were chosen to be two correlating parameters. It correlates the IFT to the density difference between gas phase and liquid phase with an exponent of 4 based on the assumption from the van der Waals equation,

$$\frac{\sigma_{hw}^{0.25}}{\rho_w - \rho_h}\left(\frac{T_{oR}}{T_c}\right)^{0.3125} = f(\Delta\rho) \tag{4.3}$$

where σ_{hw} is IFT between hydrocarbon and water, dynes/cm; ρ_w is pure water density, g/cm^3; ρ_h is density of hydrocarbon, g/cm^3; T_c is critical temperature of water, $^\circ$R; and T_{oR} is temperature, $^\circ$R. One can plot the LHS of Eq. (4.3) with respect to phase-density difference to find out their proper relationship. Danesh (1998) presented a modified version of Eq. (4.3) for modeling IFT of hydrocarbon/water systems as,

$$\sigma_{hw} = 111(\rho_w - \rho_h)^{1.024}\left(\frac{T_{oR}}{T_c}\right)^{-1.25} \tag{4.4}$$

where σ_{hw} is IFT between hydrocarbon and water, dynes/cm; ρ_w is pure water density, g/cm^3; ρ_h is density of hydrocarbon, g/cm^3; T_c is critical temperature, K; and T_{oR} is temperature, $^\circ$R.

Sutton (2009) developed another empirical correlation for determining IFT of hydrocarbon-gas/water systems,

$$\sigma_{gw} = \left[\frac{1.53988\left(\rho_{w_{g/\alpha}} - \rho_{h_{g/\alpha}}\right) + 2.08339}{\left(\frac{T_{oR}}{T_c}\right)^{\left(0.821976 - 1.83785\times10^{-3}T_{oR} + 1.34016\times10^{-6}T_{oR}^2\right)}}\right]^{3.6667} \tag{4.5}$$

where σ_{gw} is IFT between gas and water, dynes/cm; $\rho_{w_{g/\alpha}}$ is density of water phase, g/cm^3; and $\rho_{h_{g/\alpha}}$ is density of hydrocarbon-gas phase, g/cm^3; T_{oR} is temperature, $^\circ$R; T_c is critical temperature of water, $^\circ$R. As pointed out by Chalbaud et al. (2010), Firoozabadi and Ramey's correlation (Firoozabadi and Ramey Jr., 1988) might not be applicable to some gases, such as CO$_2$, because the gas solubility in water can be large.

All of the correlations mentioned above were developed based on the IFT measurements and the phase-density difference between hydrocarbon gases with water. From the experimental results, the IFT of gas mixtures with water has a strong correlation with gas composition in addition to the effect of temperature, pressure, and density difference. Considering these factors, we present a new IFT correlation,

$$
\left(\frac{\sum_{i=1}^{n} z_i M_i}{M_H} \right)^{0.183361} \frac{\sigma_{gw}^{0.25}}{\left(\rho_M^L \sum_{i=1}^{n} x_i P_i - \rho_M^V \sum_{i=1}^{n} y_i P_i \right)} T_r^{-0.3125}
$$

$$
= f \left(\rho_M^L \sum_{i=1}^{n} x_i P_i - \rho_M^V \sum_{i=1}^{n} y_i P_i \right) \tag{4.6}
$$

where z_i is the overall mole fraction of component i in the gas phase; M_i is molecular weight of component i, g/mol; M_H is the molecular weight of the heaviest component in the gas mixture, g/mol; T_r is reduced temperature of water; ρ_M^L is molar density of liquid phase, mol/m^3; ρ_M^V is molar density of vapor phase, mol/m^3; x_i is mole fraction of component i in liquid phase; and y_i is mole fraction of component i in vapor phase. This correlation takes into account the effects of pressure, temperature, individual compound's molecular weight, density difference, and gas composition on IFT of gas-mixtures/water systems.

Also, at the same temperature and pressure, a different IFT can be found for a given gas mixture due to different water salinities. Argaud (1992) presented a comprehensive review on salt's effect on IFT. Many scholars, such as Chalbaud et al. (2006), Argaud (1992), and Massoudi and King (1975), have found that there exists a unique linear relationship between IFT increment and the salt concentration of NaCl; and the slope of this line is independent of temperature when a IFT plateau is reached. Analogous to previous works (such as Standing, 1951), a linear relationship between IFT increment for CO$_2$/brine and CH$_4$/brine systems and the salt (NaCl) concentration is also provided as follows:

$$
\sigma_{cor} = k C_s \tag{4.7}
$$

where σ_{cor} represents the increase in IFT due to salinity effect, mN/m; and C_s represents the molar concentration of salt in water, mol/kg; and k is regression constant.

In this section, the salt (NaCl) effects on IFT of the CO_2/brine and CH_4/brine systems are different from each other, although a linear relationship holds between IFT increment and NaCl concentration for both systems. The IFT of a given CO_2/CH_4/brine system can be determined by first calculating the IFT for CO_2/CH_4/water systems, and then applying the following correction:

$$\sigma_{gb} = \sigma_{gw} + \gamma_{CH_4}\sigma_{cor-CH_4} + \gamma_{CO_2}\sigma_{cor-CO_2} \qquad (4.8)$$

where σ_{gb} represents IFT between gas and brine; σ_{gw} represents IFT between gas and pure water; σ_{cor-CH_4} represents the increase in IFT due to salinity effect for CH_4/brine system, mN/m; σ_{cor-CO_2} represents the increase in IFT due to salinity effect for CO_2/brine system, mN/m; γ_{CH_4} is the mole fraction of CH_4 in the original gas mixture; and γ_{CO_2} is the mole fraction of CO_2 in the original gas mixture.

4.3 Effect of pressure, temperature and salinity on IFT

In this section, the IFT measured for CO_2/brine (0–200,000 ppm NaCl) (Fig. 4.3) and CH_4/brine (0–200,000 ppm NaCl) (Fig. 4.4) at around 78.0°F, 167.0°F, and 257.0°F, respectively, is used to analyze the effect of pressure, temperature and salinity on IFT. Fig. 4.3 presents the IFT isotherms of CO_2/brine system. It indicates that, at low pressures (below around 580.2–725.2 psia), IFTs decrease approximately linearly with increasing pressure at these three temperatures, corresponding to the so-called Henry regime (Chiquet et al., 2007). Passing the Henry regime, pressure increase has less effect on the IFT reduction. When pressure increases to a high value, IFT levels off. In general, IFTs for CH_4/brine system are found to decrease with increasing temperature as shown in Fig. 4.4. On the contrary, IFTs for CO_2/brine systems measured at higher temperatures are generally higher than those measured at lower temperatures. It is because the solubility of CO_2 in water varies significantly with temperature (Yang et al., 2005). At a higher temperature, the solubility of CO_2 in water or brine is less than that at a lower temperature (Duan and Rui, 2003; Bando et al., 2003; Koschel et al., 2006). As salinity increases, the solubility of CO_2 in brine decreases, leading to changes in the brine density and IFT. As seen

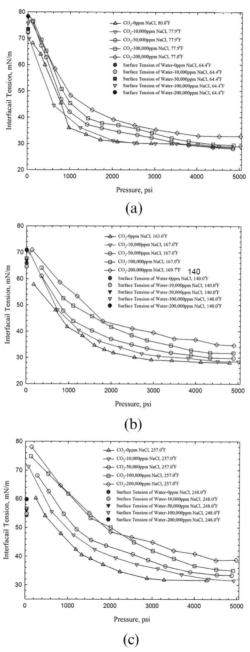

Fig. 4.3 IFT of CO_2/brine system as a function of pressure at different temperatures and different salinities. The surface tension of brine was measured by Abramzon and Gaukhberg (1993).

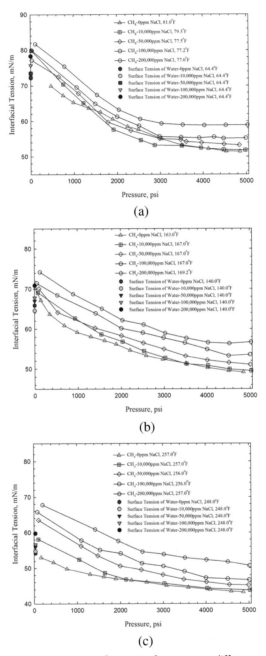

Fig. 4.4 IFT of CH_4/brine system as a function of pressure at different temperatures and different salinities. The surface tension of brine was measured by Abramzon and Gaukhberg (1993).

from Fig. 4.3, the IFT of CO_2/brine systems exhibits a more pronounced reduction with an increase in pressure at a lower temperature than that at a higher temperature. The plateau for CO_2/brine system is reached at about 1400 psia at 78.0°F, about 2000 psia at 167.0°F, and about 2900 psia at 257.0°F, respectively.

The salinity of brine in shale formations can be quite high, up to 300,000 ppm (Whitson and Brulé, 2000). Salts can affect the interfacial tension between gas and water. When ions dissolve into liquid water, electrostatic force from ions can change the original structure of water, usually forming water molecular layer around ions which is called "hydration." Indeed, water will always stride to maintain its hydrogen-bonded structure in order to maintain thermodynamic stability, while salts can affect such bonded structure formed by water, and thus affect the IFT between gas and water. In this study, IFT measurements are conducted at salinities up to 200,000 ppm of NaCl for both CO_2/brine and CH_4/brine systems, as shown in Figs. 4.3 and 4.4. Similar to CH_4/H_2O and CO_2/H_2O systems, the IFT of CH_4/brine and CO_2/brine systems exhibits a decreasing trend with increasing pressure at a given temperature. Furthermore, at the same temperature, IFT increases as more NaCl is present in water. This is attributed to the fact that, as salinity increases, the specific gravity of brine also increases; this enlarges the density difference between gas phase and liquid phase, leading to a higher IFT of the gas/brine system. At lower pressures, the IFTs corresponding to different salinities usually cross with each other in the range of 15.0–725.2 psia. Such crossing behavior might be related to the complex gas solubility in liquid phase at different temperatures (Yang et al., 2005). Chalbaud et al. (2006) presented when NaCl concentration is lower than 5000 ppm, salinity effect on IFT is negligible. For CO_2/brine systems at a low pressure, salinity shows a more obvious effect on IFT, while the salt effect on IFT reaches to a given value as the pressure becomes higher. Meanwhile, at high pressures, the average IFT increment for different NaCl concentrations depends on neither pressure nor temperature. Some scholars (Wiebe, 1941; Malinin and Kurovskaya, 1975) measured the solubility of CO_2 in water and brine as a function of pressure. They attributed the existence of a plateau to the solubility effect on the IFT reduction. Meanwhile, some studies (Wiebe, 1941; Malinin and Kurovskaya, 1975; Malinin and Savelyeva, 1972) presented that the pressure dependence of CO_2 solubility in brine exhibits a similar trend with that in pure water.

These measurements were limited to the conditions of 77.0–257.0°F and 15–5050 psia. Wiegand and Franck (1994) measured the IFT of various gases and water systems covering much greater temperature and pressure ranges. They found the IFT levels off when pressure exceeds 2175.6–2900.8 psia, while it increases very slowly with pressures above 7251.9–14,503.8 psia.

Regarding the IFT results of the CH_4/brine system as shown in Fig. 4.4, similar conclusions can be made. The CH_4/brine IFT also decreases with an increasing pressure until it reaches at a plateau. At the same temperature, CH_4/brine system needs a higher pressure to reach the plateau compared to CO_2/brine system. At about 81.0°F, for example, the value of the plateau is reached at about 2800 psia, and about 4000 psia at 163.0°F. Figs. 4.3 and 4.4 indicate that the CH_4/brine IFT is overall higher than that of CO_2/brine. The physical properties of CH_4 and CO_2 cause such difference; CH_4 has a lower solubility in water compared with CO_2 at the same temperature and pressure. Also, at the same pressure and temperature, the density difference between gas phase and liquid phase of the CH_4/brine system is larger than that of the CO_2/brine system.

As for the IFT between gas and brine, when pressure is low enough or equal to the saturation pressure of the aqueous phase, the brine/gas IFT data should approach the surface tension of water at zero pressure and the temperature of interest. The surface tensions of ordinary brine at different temperatures were measured by Abramzon and Gaukhberg (1993); these surface tensions have been labeled in Figs. 4.3 and 4.4.

For the experimental surface tension of salt solutions measured by Abramzon and Gaukhberg (1993), it is assumed that the surface tensions were measured at the saturation pressure of a specific salt solution because no specific operating pressures were reported in the paper. From Figs. 4.3 and 4.4, this constraint is only satisfied at 77.0°F. However, at 167.0°F and at atmospheric pressure, the experimental IFTs tend to be higher than the brine surface tension. This can be explained as follows. At 77.0°F, the brine drop in the IFT cell can maintain as a single liquid phase because water's saturation pressure at 77.0°F is lower than the atmospheric pressure. Therefore, the reported gas-brine IFTs data should tend to be the exact brine surface tension at this temperature. However, at higher temperatures (such as 167.0°F), water molecules are more apt to escape form liquid phase into vapor phase, which leads to a higher salinity of the brine drop and also causes a larger density difference between the liquid phase and vapor phase. Hence, relatively higher IFTs could be resulted at higher temperatures.

When the temperature is higher than the saturation temperature at atmospheric pressure, such as 257.0°F, the drop cannot maintain as a single liquid phase, but vapor phase. In the literature, surface tensions of brine or pure water reported at higher temperatures mostly were measured using the differential maximum bubble pressure method (Chen et al., 2008). For this method, a bubble chamber unit is applied, leading to a curved interface between gas and liquid phases. There may be a permanent state of metastability because of the negative pressure effect when the gas-liquid interface is curved (Firoozabadi, 2016). As shown in Figs. 4.3C and 4.4C, the deviation persists at higher temperatures; this may arise from the effect of water vaporization, as mentioned above.

4.4 Effect of CO_2 concentration on IFT

Supercritical CO_2 can be used as hydraulic fracturing fluid or enhanced gas recovery medium in shale reservoirs. Investigation on the CO_2 addition on IFT of CH_4/brine system is important for understanding the multiphase fluid flow within both the fracture and matrix. Figs. 4.5 and 4.6 show the measurement results at different temperatures. It can be seen that the presence of CO_2 in CH_4 leads to reduction in IFT between gas mixtures and brine. A lower IFT can be expected if CO_2 is added into the gas phase, but the degree of IFT reduction depends on the amount of CO_2 added. With more CO_2 present in the gas mixture, the IFT reduction effect is more pronounced. As shown in Fig. 4.5B, the IFT reduction ratio is more than 25% for the CH_4/H_2O system with 44.87 mol% CO_2 added at 163.0°F. Similarly, in Fig. 4.6A and F, the IFT reduction for the CH_4/brine systems is pronounced even at low concentrations of CO_2. The density difference between gas phase and liquid phase is reduced if CO_2 is added to CH_4, which is a major factor causing IFT reduction. Another reason is because CO_2 exhibits a higher solubility in water or brine compared to CH_4, further decreasing the density difference. The phase behavior of CO_2/CH_4 mixture with water or brine, together with the physical properties of the gas components, all contribute to the IFT reduction effect.

Rushing et al. (2008) measured the effect of CO_2 concentration (up to 20.00 mol%) on IFT of gas-water system (gas: CH_4 with a small fraction of C_2H_6 and C_3H_8) at high pressure-temperature conditions. They suggested

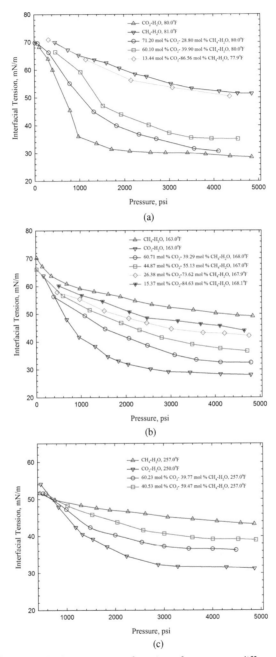

Fig. 4.5 IFT of $CO_2/CH_4/H_2O$ systems as a function of pressure at different temperatures.

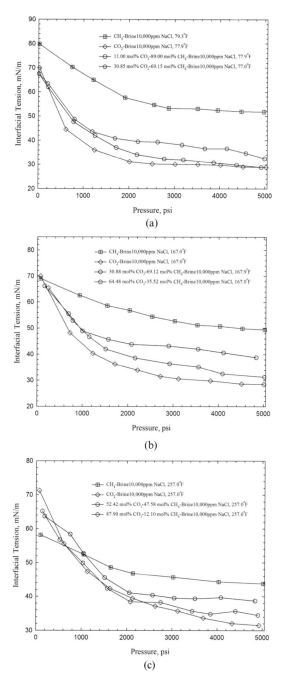

Fig. 4.6 IFT of CO_2/CH_4/brine as a function of pressure at different temperatures and NaCl concentrations.

(Continued)

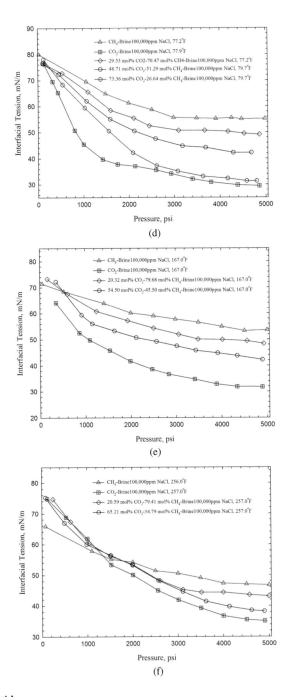

Fig. 4.6, cont'd

(Continued)

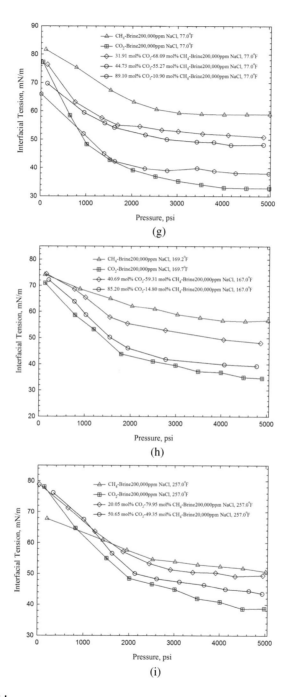

Fig. 4.6, cont'd

that a higher concentration of CO_2 resulted in a lower IFT over a much greater pressure range than that for gases with lower CO_2 concentrations. They found that CO_2 concentration in vapor phase tended to decrease IFT of CH_4/brine systems at lower pressures, but slightly increased the IFT or showed no effect at higher pressures. It is, however, shown in our study that the presence of CO_2 has a significant effect on the IFT at both high and low pressures. Shariat (2014) measured IFT for gas mixtures containing up to 20.00 mol% CO_2 (or without CO_2) with water over a wider pressure range, showing that the presence of CO_2 up to 20.00 mol% in gas mixtures has no significant effect on gas–water IFTs at higher pressures. Ren et al. (2000) measured the IFT of CO_2/CH_4/H_2O systems with CO_2 concentrations of 0, 20.00, 40.00, 60.00, and 80.00 mol% at temperatures of 104.0°F, 140.0°F, 176.0°F, and 212.0°F, respectively. They found that: CO_2 concentration of 20.00 mol% leads to negligible IFT reduction, and CO_2 concentrations of 20.00–40.00 mol% lead to minor reduction in IFT. They also reported IFT reduction at higher concentrations of CO_2 in gas mixture for all temperatures. However, we observed pronounced IFT reduction even at low CO_2 concentrations in our study.

The aforementioned experimental results demonstrate that CO_2 decreases the IFT of CH_4/H_2O systems, while salinity tends to increase the IFT of CH_4/H_2O systems. These IFT data are useful for assessing the engineering soundness of either using CO_2 for fracturing shale formations or CO_2 huff-and-puff for enhancing shale gas recovery. For a given shale reservoir, if the reservoir conditions such as reservoir temperature, pressure and salinity of formation water are given, the IFT between shale gas (mainly CH_4) and brine can be approximately determined.

Taking reservoir conditions, 167.0°F and 4351 psia, for example, the IFT between shale gas (mainly CH_4) with brine (with a salinity of 100,000 ppm NaCl) is about 54.50 mN/m, about 4.00 mN/m higher than that of CH_4/H_2O system (see Fig. 4.7). IFT significantly affects the in situ capillary pressure and entrapment of gas in shale matrix; in order to enhance the shale gas recovery, gas/water IFT should be as low as possible. To reduce the IFT of CH_4/brine (with a salinity of 100,000 ppm NaCl) system to a value of 45.50 mN/m, CO_2 concentration in the gas has to be around 50.00 mol%. Similarly, if the salinity of reservoir brine is 200,000 ppm, the concentration of CO_2 should be around 20.05–50.65 mol% to obtain the same level of IFT at the NaCl concentration of 50,000 ppm at the pressure of 4351 psia and temperature of 257.0°F, as shown in Fig. 4.8.

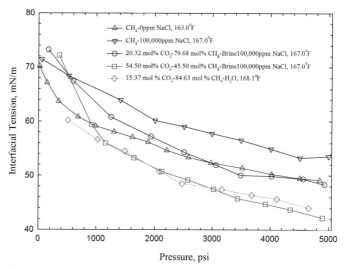

Fig. 4.7 IFT of CH_4/CO_2/brine systems as a function of pressure.

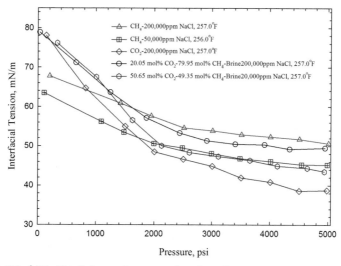

Fig. 4.8 IFT of CO_2/CH_4/brine systems as a function of pressure.

4.5 IFT modeling for CO_2/CH_4/H_2O and CO_2/CH_4/brine systems

4.5.1 Improved IFT model for CO_2/CH_4/H_2O systems

The IFT data for CO_2/CH_4/H_2O systems are used to regress the coefficients appearing in Eq. (4.6). The following correlation is obtained based on regression analysis as indicated by Fig. 4.9 ($R^2 = 0.9658$),

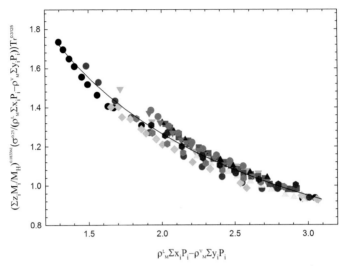

Fig. 4.9 Regression of IFT model parameters using measured data in this study for $CO_2/CH_4/H_2O$ systems: ●, CO_2/H_2O, 80.0°F; ●, CO_2/H_2O, 163.0°F; ▼, CO_2/H_2O, 250.0°F; ▲, CH_4/H_2O, 81.0°F; ■, CH_4/H_2O, 163.0°F; ■, CH_4/H_2O, 257.0°F; ◆, 60.10 mol% $CO_2/39.90$ mol% CH_4/H_2O, 80.0°F; ◆, 71.20 mol% $CO_2/28.80$ mol% CH_4/H_2O, 80.0°F; ▲, 26.40 mol% $CO_2/73.60$ mol% CH_4/H_2O, 168.0°F; ▼, 15.40 mol% $CO_2/84.60$ mol% CH_4/H_2O, 168.0°F; ●, 60.70 mol% $CO_2/39.30$ mol% CH_4/H_2O, 168.0°F; ●, 44.90 mol% $CO_2/55.10$ mol% CH_4/H_2O, 167.0°F; ●, 13.40 mol% $CO_2/86.60$ mol% CH_4/H_2O, 78.0°F; ●, 40.50 mol% $CO_2/59.50$ mol% CH_4/H_2O, 257.0°F; ▼, 60.20 mol% $CO_2/39.80$ mol% CH_4/H_2O, 257.0°F.

$$\sigma_{gw}^{0.25} = \frac{2.068}{T_r^{0.3125}} \left(\frac{M_H}{\sum_{i=1}^{n} z_i M_i} \right)^{0.183361} \left(\rho_M^L \sum_{i=1}^{n} x_i P_i - \rho_M^V \sum_{i=1}^{n} y_i P_i \right)^{0.2921} \tag{4.9}$$

Fig. 4.10 presents a parity chart that plots the measured IFTs versus the calculated ones with Eq. (4.9), while Fig. 4.11 shows the distribution of errors versus the number of data points. Both figures illustrate that a good match is obtained between the measured and calculated IFTs, demonstrating the accuracy of Eq. (4.9) in correlating the IFTs for $CO_2/CH_4/H_2O$ systems. Fig. 4.11 indicates that the difference between predicted and measured data is mostly less than ±5.00 mN/m. Table 4.3 summarizes the statistical analysis

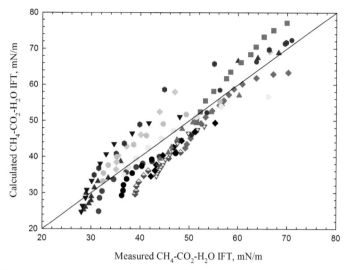

Fig. 4.10 Comparison between predicted IFTs with Eq. (4.9) versus measured IFTs for CO$_2$/CH$_4$/H$_2$O systems: ●, CO$_2$/H$_2$O, 80.0°F; ●, CO$_2$/H$_2$O, 163.0°F; ▼, CO$_2$/H$_2$O, 250.0°F; ▲, CH$_4$/H$_2$O, 81.0°F; ■, CH$_4$/H$_2$O, 163.0°F; ■, CH$_4$/H$_2$O, 257.0°F; ◆, 60.10 mol% CO$_2$/39.90 mol% CH$_4$/H$_2$O, 80.0°F; ◆, 71.20 mol% CO$_2$/28.80 mol% CH$_4$/H$_2$O, 80.0°F; ▲, 26.40 mol% CO$_2$/73.60 mol% CH$_4$/H$_2$O, 168.0°F; ▼, 15.40 mol% CO$_2$/84.60 mol% CH$_4$/H$_2$O, 168.0°F; ●, 60.70 mol% CO$_2$/39.30 mol% CH$_4$/H$_2$O, 168.0°F; ●, 44.90 mol% CO$_2$/55.10 mol% CH$_4$/H$_2$O, 167.0°F; ●, 13.40 mol% CO$_2$/86.60 mol% CH$_4$/H$_2$O, 78.0°F; ●, 40.50 mol% CO$_2$/59.50 mol% CH$_4$/H$_2$O, 257.0°F; ▼, 60.20 mol% CO$_2$/39.80 mol% CH$_4$/H$_2$O, 257.0°F.

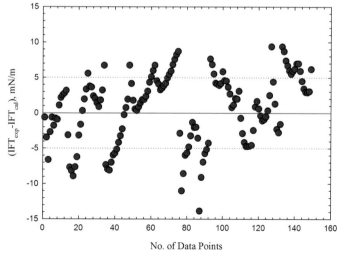

Fig. 4.11 Difference between measured IFTs and predicted IFT with Eq. (4.9) for CO$_2$/CH$_4$/H$_2$O systems.

Table 4.3 Performance of different IFT models in reproducing the IFTs of $CO_2/CH_4/H_2O$ systems.

Model	Number of data points obtained by this study	AARE (%)	SD (%)
New Model	156	9.42	11.33
Danesh (1998)	156	28.28	38.48
Sutton (2009)	156	23.13	40.39
Firoozabadi and Ramey Jr. (1988)	156	25.52	29.08
Weinaug and Katz (1943)	156	35.98	44.11

results on using the improved correlation. As shown in Table 4.3, the average absolute relative error (AARE) and standard deviation (SD) are found to be 9.42% and 11.33%, respectively.

4.5.2 Comparison with existing correlations

The improved model is compared with four commonly used hydrocarbon gas/H_2O IFT correlations in the literature. As for the Firoozabadi and Ramey Jr. (1988) correlation, a relationship between $\frac{\sigma_{gw}^{0.25}}{(\rho_L-\rho_V)}T_r^{0.3125}$ and $(\rho_L-\rho_V)$ can be obtained by plotting these two terms together, and then applying proper regression, as done in Fig. 4.12. The following equation is obtained,

$$\sigma_{gw}^{0.25} = \frac{2.978}{T_r^{0.3125}}(\rho_L - \rho_V)^{0.1382} \qquad (4.10)$$

Fig. 4.13 shows a parity chart that compares the measured and calculated IFTs with Eq. (4.10). Similarly, three commonly-used correlations (Sutton, 2009; Danesh, 1998; Weinaug and Katz, 1943) are also used to correlate the IFT data, as shown in Figs. 4.14–4.16. Table 4.3 summarizes the overall comparison results. As can be seen from Figs. 4.13–4.16 and Table 4.3, there is a large discrepancy between the measured IFTs and calculated IFTs with these four models. Compared with these four correlations, Eq. (4.9) provides more accurate IFT prediction for the $CO_2/CH_4/H_2O$ systems.

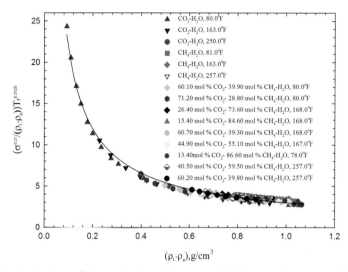

Fig. 4.12 Application of Firoozabadi and Ramey Jr. (1988) correlation to the IFT data of $CO_2/CH_4/H_2O$ systems.

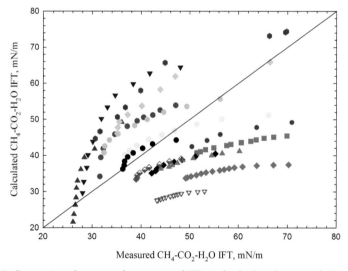

Fig. 4.13 Comparison between the measured IFTs and calculated ones with Firoozabadi and Ramey Jr. (1988) correlation for $CO_2/CH_4/H_2O$ systems.

4.5.3 Validation of the improved model

In order to further test the predictive ability of the newly developed IFT model, Eq. (4.9) is tested with 150 IFT data presented by Ren et al. (2000) that are not used in developing the new IFT model. The comparison

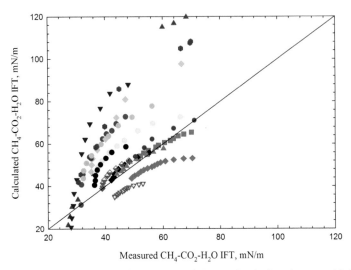

Fig. 4.14 Comparison between the measured IFTs and calculated ones with Danesh (1998) correlation for $CO_2/CH_4/H_2O$ systems.

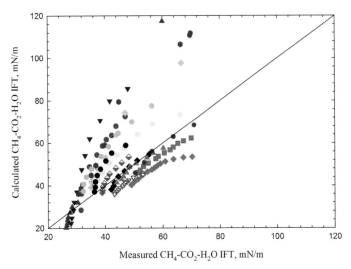

Fig. 4.15 Comparison between the measured IFTs and calculated ones with Sutton (2009) correlation for $CO_2/CH_4/H_2O$ systems.

between the calculated and measured IFT values is presented in Fig. 4.17. It indicates that the new IFT model provides a good prediction on $CO_2/CH_4/H_2O$ IFT. Deviations between the measured and predicted IFTs are mostly less than $\pm 5.00\,mN/m$. Table 4.4 shows the validation results. Compared to

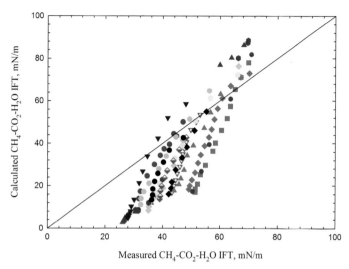

Fig. 4.16 Comparison between the measured IFTs and predicted ones with Weinaug and Katz (1943) correlation for CO$_2$/CH$_4$/H$_2$O systems.

other models, the improved model provides more accurate IFT prediction for CO$_2$/CH$_4$/H$_2$O systems, with AARE and SD of 10.19% and 14.01%, respectively.

4.5.4 IFT modeling for CO$_2$/CH$_4$/brine systems

The preceding discussion shows that salinity can increase the IFT of CO$_2$/CH$_4$/H$_2$O systems. Some researchers (Chalbaud et al., 2006; Argaud, 1992) proposed to use a linear relationship to account for the IFT increase as a function of salinity. It can be seen from Figs. 4.3 and 4.4 that, after the IFT levels off, the IFT increase due to salinity effect tends to be independent of temperature and pressure. As for the CO$_2$/brine (NaCl) systems, the following linear relationship is found to be adequate to account for the salinity effect, as shown in Fig. 4.18:

$$\sigma_{cor-\mathrm{CO}_2} = 3.45 \times 10^{-5} C_S \qquad (4.11)$$

As for the CH$_4$/brine (NaCl) systems, the following linear relationship is obtained, as shown in Fig. 4.18:

$$\sigma_{cor-\mathrm{CH}_4} = 3.45 \times 10^{-5} C_S \qquad (4.12)$$

The ratio of 3.600×10^{-5} obtained for CO$_2$/brine (NaCl) system is slightly different from the values reported in the literature. Chalbaud et al. (2006)

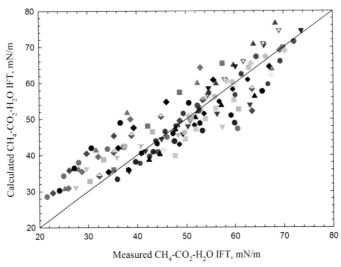

Fig. 4.17 Comparison between the measured data by Ren et al. and predicted IFTs with the new correlation for CO_2/CH_4/H_2O systems: ▲, CH_4/H_2O, 77.0°F; ▼, 20.00 mol% CO_2/80.00 mol% CH_4/H_2O, 77.0°F; ●, 40.00 mol% CO_2/60.00 mol% CH_4/H_2O, 77.0°F; ■, 60.00 mol% CO_2/40.00 mol% CH_4/H_2O, 77.0°F; ◆, 80.00 mol% CO_2/20.00 mol% CH_4/H_2O, 77.0°F; ▽,CH_4/H_2O, 104.0°F; ◇, 20.00 mol% CO_2/80.00 mol% CH_4/H_2O, 104.0°F; ●, 40.00 mol% CO_2/60.00 mol% CH_4/H_2O, 104.0°F; ◆, 60.00 mol% CO_2/40.00 mol% CH_4/H_2O, 104.0°F; ▲, 80.00 mol% CO_2/20.00 mol% CH_4/H_2O, 104.0°F; ●, CH_4/H_2O, 140.0°F; ◌, 20.00 mol% CO_2/80.00 mol% CH_4/H_2O, 140.0°F; ●, 40.00 mol% CO_2/60.00 mol% CH_4/H_2O, 140.0°F; ◆, 60.00 mol% CO_2/40.00 mol% CH_4/H_2O, 140.0°F; ●, 80.00 mol% CO_2/20.00 mol% CH_4/H_2O, 140.0°F; ✚, CH_4/H_2O, 176.0°F; ▲, 20.00 mol% CO_2/80.00 mol % CH_4/H_2O, 176.0°F; ▼, 40.00 mol% CO_2/60.00 mol% CH_4/H_2O, 176.0°F; ■, 60.00 mol% CO_2/40.00 mol% CH_4/H_2O, 176.0°F; ◆, 80.00 mol% CO_2/20.00 mol% CH_4/H_2O, 176.0°F; ■, CH_4/H_2O, 212.0°F; ●, 20.00 mol% CO_2/80.00 mol% CH_4/H_2O, 212.0°F; ●, 40.00 mol% CO_2/60.00 mol% CH_4/H_2O, 212.0°F; ▼, 60.00 mol% CO_2/40.00 mol% CH_4/H_2O, 212.0°F; ●, 80.00 mol% CO_2/20.00 mol% CH_4/H_2O, 212.0°F.

Table 4.4 Performance of the improved IFT model in predicting the IFTs of CO_2/CH_4/H_2O systems published by Ren et al. (2000).

Model	Number of data points measured by Ren et al.	AARE (%)	SD (%)
This study	150	10.19	14.01
Danesh (1998)	150	57.92	62.82
Sutton (2009)	150	45.83	50.08
Firoozabadi and Ramey Jr. (1988)	150	26.56	30.40
Weinaug and Katz (1943)	150	26.45	32.67

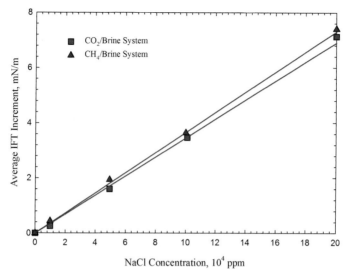

Fig. 4.18 Average IFT increment for CO_2/brine and CH_4/brine systems as a function of NaCl concentration.

reported a ratio of 2.550×10^{-5} instead of 3.600×10^{-5} for the CO_2/brine system (up to 16,071 ppm of NaCl). Massoudi and King (1975) reported a ratio of 2.704×10^{-5}, while Argaud (1992) obtained a ratio of 2.789×10^{-5} for the CO_2/brine system. The ratio of 3.950×10^{-5} obtained in this study for CH_4/brine systems is larger than that of CO_2/brine systems. It indicates that an increase in salinity (NaCl) results in a more pronounced increase in the IFT for CH_4/brine systems than for CO_2/brine systems. Cai et al. (1996) measured IFT of salt solutions containing NaCl, $CaCl_2$ and $MgCl_2$ with n-C_8H_{18}. They showed that the degree of IFT increase is sensitive to salt species. For salts such as KCl, $CaCl_2$ and $MgCl_2$, the effect of salt on IFT has been widely studied, and such increase in IFT is not linear at salt concentrations higher than 1.0 mol/kg. It is noted that one can first calculate the IFT of CO_2/CH_4/H_2O system with Eq. (4.9), and then apply Eqs. (4.8), (4.11), and (4.12) to obtain the IFT of a given CO_2/CH_4/brine (NaCl) system.

References

Abramzon, A.A., Gaukhberg, R.D., 1993. Surface tension of salt solutions. Russ. J. Appl. Chem. 66, 1428.

Aggelopoulos, C.A., Robin, M., Perfetti, M., Vizika, O., 2010. CO_2/$CaCl_2$ solution interfacial tensions under CO_2 geological storage conditions: influence of cation valence on interfacial tension. Adv. Water Resour. 33, 691.

Aggelopoulos, C.A., Robin, M., Vizika, O., 2011. Interfacial tension between CO_2 and brine ($NaCl + CaCl_2$) at elevated pressures and temperatures: the additive effect of different salts. Adv. Water Resour. 34, 505.

Akutsu, T., Yoshinori, Y., Hirokazu, Y., Masaru, W., Richard, L.S.J., Hiroshi, I., 2007. Interfacial tension between water and high pressure CO_2 in the presence of hydrocarbon surfactants. Fluid Phase Equilib. 257, 163.

Almeida, B.S., Telo da Gama, M.M., 1989. Surface tension of simple mixtures: comparison between theory and experiment. J. Phys. Chem. 93, 4132.

Argaud, M.J., 1992. Predicting the interfacial tension of brine/gas (or condensate) systems. In: Presented at the SCA European Core Analysis Symposium. Paris.

Ayirala, S.C., Rao, D.N., 2004. Application of a new mechanistic parachor model to predict dynamic gas-oil miscibility in reservoir crude oil-solvent systems. In: Presented at the SPE International Petroleum Conference. Puebla. SPE 91920.

Ayirala, S.C., Rao, D.N., 2006. A new mechanistic parachor model to predict dynamic interfacial tension and miscibility in multicomponent hydrocarbon systems. J. Colloid Interface Sci. 299, 321.

Bachu, S., Bennion, D.B., 2009. Interfacial tension between CO_2, freshwater, and brine in the range of pressure from (2 to 27) MPa, temperature from (20 to 125) °C, and water salinity from (0 to 334 000) $mg \cdot L^{-1}$. J. Chem. Eng. Data 54, 765.

Bahramian, A., Danesh, A., 2004. Prediction of liquid-liquid interfacial tension in multicomponent systems. Fluid Phase Equilib. 221, 197.

Bahramian, A., Danesh, A., 2005. Prediction of liquid-vapour surface tension in multicomponent systems. Fluid Phase Equilib. 236, 156.

Bando, S., Takemura, F., Nishio, M., Hihara, E., Akai, M., 2003. Solubility of CO_2 in aqueous solutions of NaCl at (30 to 60) °C and (10 to 20) MPa. J. Chem. Eng. Data 48, 576.

Bennion, D.B., Bachu, S., 2008. A correlation of the interfacial tension between supercritical phase co_2 and equilibrium brine as a function of salinity, temperature and pressure. In: Presented at the SPE Annual Technical Conference and Exhibition. Colorado. SPE 114479.

Bongiorno, V., Davis, H.T., 1975. Modified van der Waals theory of fluid interfaces. Phys. Rev. A 12, 2213.

Cahn, J.W., Hilliard, J.E., 1958. Free energy of a nonuniform system. I. Interfacial free energy. J. Chem. Phys. 28, 258.

Cai, B.Y., Yang, J.T., Guo, T.M., 1996. Interfacial tension of hydrocarbon + water/brines systems under high pressure. J. Chem. Eng. Data 41, 493.

Chalbaud, C., Robin, M., Egermann, P., 2006. Interfacial tension data and correlations of brine/CO_2 systems under reservoir conditions. In: Presented at the SPE Annual Technical Conference and Exhibition, San Antonio. SPE 102918.

Chalbaud, C., Robin, M., Lombard, J.M., Martin, F., Egermann, P., Bertin, H., 2009. Interfacial tension measurement and wettability evaluation for geological CO_2 storage. Adv. Water Resour. 32, 98.

Chalbaud, C., Robin, M., Lombard, J.M., Bertin, H., Egermann, P., 2010. Brine/CO_2 interfacial properties and effects on CO_2 storage in deep saline aquifers. Oil Gas Sci. Technol. 65, 541.

Chen, Z., Xia, S., Ma, P., 2008. Measuring surface tension of liquids at high temperature and elevated pressure. J. Chem. Eng. Data 53, 742.

Chiquet, P., Daridon, J., Broseta, D., Thibeau, S., 2007. CO_2/water interfacial tensions under pressure and temperature conditions of CO_2 geological storage. Energ. Conver. Manage. 48, 736.

Chun, B.S., Wilkinson, G.T., 1995. Interfacial tension in high-pressure carbon dioxide mixtures. Ind. Eng. Chem. Res. 34, 4371.

da Rocha, S.R.P., Harrison, K.L., Johnston, K.P., 1999. Effect of surfactants on the interfacial tension and emulsion formation between water and carbon dioxide. Langmuir 15, 419.

Danesh, A., 1998. PVT and Phase Behaviour of Petroleum Reservoir Fluids (Ph.D. dissertation). Herriot Watt University, Edinburgh.

Duan, Z., Rui, S., 2003. An improved model calculating CO$_2$ solubility in pure water and aqueous NaCl solutions from 273 to 533 K and from 0 to 2000 bar. Chem. Geol. 193, 257.

Evans, R., 1979. The nature of the liquid-vapor Interface and other topics in the statistical mechanics of non-uniform, classical fluids. Adv. Phys. 28, 143.

Fawcett, M.J., 1994. Evaluation of correlations and parachors to predict low interfacial tensions in condensate systems. In: Presented at the SPE 69th Annual Technical Conference and Exhibition. New Orleans. SPE 28611.

Firoozabadi, A., 2016. Thermodynamics and Applications in Hydrocarbon Energy Production. Mc Graw Hill Education, New York.

Firoozabadi, A., Ramey Jr., H.J., 1988. Surface tension of water-hydrocarbon systems at reservoir conditions. J. Can. Pet. Technol. 27, 41.

Georgiadis, A., Maitland, G., Trusler, J.P.M., Bismarck, A., 2010. Interfacial tension measurements of the (H$_2$O + CO$_2$) system at elevated pressures and temperatures. J. Chem. Eng. Data 55, 4168.

Hebach, A., Oberhof, A., Dahmen, N., Kogel, A., Ederer, H., Dinjus, E., 2002. Interfacial tension at elevated pressures-measurements and correlations in the water + carbon dioxide system. J. Chem. Eng. Data 47, 1540.

Heller, R., Zoback, M., 2014. Adsorption of methane and carbon dioxide on gas shale and pure mineral samples. J. Unconv. Oil Gas Resour. 8, 14.

Heuer, G.J., 1957. Interfacial Tension of Water against Hydrocarbons and Other Gases and Adsorption of Methane on Solids at Reservoir Temperatures and Pressures (Ph.D. dissertation). University of Texas, Austin, TX.

Hocott, C.R., 1939. Interfacial tension between water and oil under reservoir conditions. Trans. AIME 132, 184.

Hough, E.W., Stegemeier, G.L., 1961. Correlation of surface and interfacial tension of light hydrocarbons in the critical region. SPE J. 1, 259.

Hough, E.W., Rzasa, M.J., Wood, B.B., 1951. Interfacial tensions at reservoir pressures and temperatures; apparatus and the water-methane systems. J. Petrol. Tech. 192, 57.

Hussen, C., Amin, R., Madden, G., Evans, B., 2012. Reservoir simulation for enhanced gas recovery: an economic evaluation. J. Nat. Gas Sci. Eng. 5, 42.

Jennings, H.Y., Newman, G.H., 1971. The effect of temperature and pressure on the interfacial tension of water against methane-normal decane mixtures. SPE J. 11, 171.

Jho, C., Nealon, D., Shogbola, S., King, A.D., 1978. Effect of pressure on the surface tension of water: adsorption of hydrocarbon gases and carbon dioxide on water at temperatures between 0 and 50°C. J. Colloid Interface Sci. 65, 141.

Johansson, K., Eriksson, J.C., 1974. γ and dγ/dT Measurements on aqueous solutions of 1,1-electrolyte. J. Colloid Interface Sci. 49, 469.

Khashefi, K., Pereira, L.M.C., Chapoy, A., Burgass, R., Tohidi, B., 2016. Measurement and modelling of interfacial tension in methane/water and methane/brine systems at reservoir conditions. Fluid Phase Equilib. 409, 301.

Khosharay, S., Varaminian, F., 2014. Experimental and modeling investigation on surface tension and surface properties of (CH$_4$+H$_2$O), (C$_2$H$_6$+H$_2$O), (CO$_2$+H$_2$O) and (C$_3$H$_8$+H$_2$O) from 284.15 K to 312.15 K and pressures up to 60 bar. Int. J. Refrig. 47, 26.

Koschel, D., Coxam, J.Y., Majer, V., 2006. Enthalpy and solubility data of CO$_2$ in water and NaCl (aq) at conditions of interest for geological sequestration. Fluid Phase Equilib. 247, 107.

Kvamme, B., Kuznetsova, T., Hebach, A., Oberhof, A., Lunde, E., 2007. Measurements and modelling of interfacial tension for water + carbon dioxide systems at elevated pressures. Comput. Mater. Sci. 38, 506.

Lee, S.T., Chien, M.C.H., 1984. A new multicomponent surface tension correlation based on scaling theory. In: Presented at the SPE/DOE Improved Oil Recovery Conference. Tulsa. SPE/DOE 12643.

Lepski, B., 1997. Gravity Assisted Tertiary Gas Injection Process in Water Drive Oil Reservoirs (Ph.D. dissertation). Louisiana State University, Baton Rouge, LA.

Levin, Y., dos Santos, A.P., Diehl, A., 2009. Ions at the air–water interface: an end to a hundred-year-old mystery? Phys. Rev. Lett. 103, 1.

Li, X., Elsworth, D., 2014. Geomechanics of CO_2 enhanced shale gas recovery. J. Nat. Gas Sci. Eng. 26, 1607.

Li, X., Yang, D., 2013. Determination of mutual solubility between CO_2 and water by using the Peng-Robinson equation of state with modified alpha function and binary interaction parameter. Ind. Eng. Chem. Res. 52, 13829.

Li, X., Boek, E., Maitland, G.C., Trusler, J.P.M., 2012a. Interfacial tension of (Brines $+ CO_2$): (0.864 NaCl $+ 0.136$ KCl) at temperatures between (298 and 448) K, pressures between (2 and 50) MPa, and total molarities of (1 to 5) mol$\cdot$kg^{-1}. J. Chem. Eng. Data 57, 1078.

Li, X., Boek, E., Maitland, G.C., Trusler, J.P.M., 2012b. Interfacial tension of (Brines $+ CO_2$): $CaCl_2$(aq), $MgCl_2$(aq), and Na_2SO_4(aq) at temperatures between (343 and 423) K, pressures between (2 and 50) MPa, and molarities of (0.5 to 5) mol$\cdot$kg^{-1}. J. Chem. Eng. Data 57, 1369.

Li, Z., Wang, S., Li, S., Liu, W., Li, B., Lv, Q.C., 2013. Accurate determination of the CO_2-brine interfacial tension using graphical alternating conditional expectation. Energy Fuel 28, 624.

Macleod, D.B., 1923. On a relation between surface tension and density. Trans. Faraday Soc. 19, 38.

Malinin, S.D., Kurovskaya, N.A., 1975. Solubility of CO_2 in chloride solutions at elevated temperatures and CO_2 pressures. Geochem. Int. 2, 199.

Malinin, S.D., Savelyeva, N.I., 1972. The solubility of CO_2 in NaCl and $CaCl_2$ solutions at 25, 50 and 75°C under elevated CO_2 pressures. Geochem. Int. 9, 410.

Massoudi, R., King, A.D., 1975. Effect of pressure on the surface tension of aqueous solutions. adsorption of hydrocarbon gases, carbon dioxide, and nitrous oxide on aqueous solutions of sodium chloride and tetra-n-butylammonium bromide at 25°C. J. Phys. Chem. 79, 1670.

Massoudi, R., King Jr., A.D., 1974. Effect of pressure on the surface tension of water adsorption of low molecular weight gases on water at 25°C. J. Phys. Chem. 78, 2262.

Nordholm, S., Johnson, M., Freasier, B.C., 1980. Generalized van der Waals theory. III. The prediction of hard sphere structure. Aust. J. Chem. 33, 2139.

Park, J.Y., Lim, J.S., Yoon, C.H., Lee, C.H., Park, K.P., 2005. Effect of a fluorinated sodium Bis (2-ethylhexyl) sulfosuccinate (aerosol-OT, AOT) analogue surfactant on the interfacial tension of $CO_2 +$ water and $CO_2 +$ Ni-plating solution in near- and supercritical CO_2. J. Chem. Eng. Data 50, 299.

Pegram, L.M., Record, M.T., 2008. The thermodynamic origin of hofmeister ion effects. J. Phys. Chem. B 112, 9428.

Pereira, L.M.C., Chapoy, A., Burgass, R., Oliveira, M.B., Coutinho, J.A.P., Tohidi, B., 2016. Study of the impact of high temperature and pressures on the equilibrium densities and interfacial tension of the carbon dioxide/water system. J. Chem. Thermodyn. 93, 404.

Quayle, O.R., 1953. The parachors of organic compounds. An interpretation and catalogue. Chem. Rev. 53, 439.

Ralston, J., Healy, T., 1973. Specific cation effects on water structure at the air–water and air-octadecanol monolayer–water interface. J. Colloid Interface Sci. 42, 629.

Ren, Q.Y., Chen, G.J., Yan, W., Guo, T.M., 2000. Interfacial tension of $(CO_2 + CH_4)$+water from 298 K to 373 K and pressures up to 30 MPa. J. Chem. Eng. Data 45, 610.

Rowlinson, J.S., 1979. Translation of J. D. van der Waals' "the thermodynamic theory of capillarity under the hypothesis of a continuous variation of density". J. Stat. Phys. 20, 197.

Rushing, J.A., Newsham, K.E., Van Fraassen, K.C., Mehta, S.A., Moore, G.R., 2008. Laboratory measurements of gas-water interfacial tension at HP/HT reservoir conditions. In: Presented at the CIPC/SPE Gas Technology Symposium. Calgary. SPE 114516.

Sachs, W., Meyn, V., 1995. Pressure and temperature dependence of the surface tension in the system natural gas/water: principles of investigation and the first precise experimental data for pure methane/water at 25°C up to 48.8 MPa. Colloids Surf. A Physicochem. Eng. Asp. 94, 291.

Sato, K., 2003. Sensitivity of interfacial-tension predictions to parachor-method parameters. J. Jpn. Pet. Inst. 46, 148.

Shah, V., Broseta, D., Mouronval, G., Montel, F., 2008. Water/acid gas interfacial tensions and their impact on acid gas geological storage. Int. J. Greenh. Gas Control 2, 594.

Shariat, A., 2014. Measurement and Modelling of Interfacial Tension at High Pressure/High Temperature Conditions (Ph.D. dissertation). University of Calgary, Calgary, AB.

Shariat, A., Moore, R.G., Mehta, S.A., Van Fraassen, K.C., Newsham, K.E., Rushing, J.A., 2011. A laboratory study of the effects of fluid compositions on gas-water interfacial tension at HP/HT reservoir conditions. In: Presented at the SPE Annual Technical Conference and Exhibition. Denver. SPE 146178.

Shariat, A., Moore, R.G., Mehta, S.A., Fraassen, K., Newsham, K., Rushing, J.A., 2012. Laboratory measurement of CO_2-H_2O interfacial tension at HP/HT conditions: implications for CO_2 sequestration in deep aquifers. In: Presented at the Carbon Management Technology Conference. Orlando. Paper 150010.

Søreide, I., Whitson, C.H., 1992. Peng-Robinson prediction for hydrocarbons, CO_2, N_2, and H_2S with pure water and NaCl brine. Fluid Phase Equilib. 77, 217.

Standing, M.B., 1951. Volumetric and Phase Behaviour of Oil Hydrocarbon Systems. California Research Corp, Dallas, TX.

Sudgen, S., 1921. Capillary rise. J. Chem. Soc. 119, 1483.

Sutjiadi-Sia, Y., Jaeger, P., Eggers, R., 2008. Interfacial phenomena of aqueous systems in dense carbon dioxide. J. Supercrit. Fluids 46, 272.

Sutton, R.P., 2009. An improved model for water-hydrocarbon surface tension at reservoir conditions. In: Presented at the SPE Annual Technical Conference and Exhibition. New Orleans. SPE 124968.

Tewes, F., Boury, F., 2005. Formation and rheological properties of the supercritical CO_2-water pure interface. J. Phys. Chem. B 109, 3990.

Tian, Y., Xiao, Y., Zhu, H., Dong, X., Ren, X., Zhang, F., 1997. Interfacial tension between water and non-polar fluids at high pressures and high temperatures. Acta Phys. -Chim. Sin. 13, 89.

Weinaug, C.F., Katz, D.L., 1943. Surface tension of methane-propane mixtures. J. Ind. Eng. Chem. 35, 239.

Whitson, C.H., Brulé, M.R., 2000. Phase Behavior. Henry L. Doherty Memorial Fund of AIME Society of Petroleum Engineers Inc, Richardson, TX.

Wiebe, R., 1941. The brine system carbon dioxide-water under pressure. Chem. Rev. 29, 475.

Wiegand, G., Franck, E.U., 1994. Interfacial tension between water and non-polar fluids up to 473 K and 2800 bar. Ber. Bunsen. Phys. Chem 98, 809.

Yan, W., Zhao, G., Chen, G., Guo, T., 2001. Interfacial tension of (methane + nitrogen) +water and (carbon dioxide + nitrogen)+water systems. J. Chem. Eng. Data 46, 1544.

Yang, D.Y., Tontiwachwuthikul, P., Gu, Y.A., 2005. Interfacial interactions between reservoir brine and CO_2 at high pressures and elevated temperatures. Energy Fuel 19, 216.

Zhao, G.Y., 2002. Measurement and calculation of high pressure interfacial tension of methane nitrogen/water system. J. Univ. Pet. 26, 75.

Zuo, Y.X., Stenby, E.H., 1997. Corresponding-states and parachor models for the calculation of interfacial tensions. Can. J. Chem. Eng. 75, 1130.

CHAPTER FIVE

Oil/gas recovery and CO$_2$ sequestration in shale

5.1 Selective adsorption of CO$_2$/CH$_4$ mixture on clay-rich shale using molecular simulations

With an increasing demand for fossil fuels, shale gas is accepted as an important substitute for conventional fossil resources, owning to its huge reserves and potential CO$_2$ sequestration (Burnham et al., 2011; Hughes, 2013; Wu et al., 2015a,b; Falk et al., 2015). Generally, shale gas storage in shale matrix exists in three patterns, i.e., adsorbed gas on rock surface (a dense and liquid-like phase), dissolved gas in organic matter, and free gas existing in pore networks (Curtis, 2002; Chen et al., 2018; Etminan et al., 2014). With the gas storage in a dense and liquid-like adsorbed state, the overall gas-storage capacity is greatly improved relative to that in a free state alone (Heller and Zoback, 2014). Based on the characteristics of gas-bearing reservoirs, the volume proportion of adsorbed gas accounts for a sizeable shale of the total reserves in five US shale formations, ranging from 20 vol% to 85 vol% (Montgomery et al., 2005). In addition, the adsorbed gas could occupy more than 50% of the total gas reserves in the Devonian shales (Lu et al., 1995). CO$_2$ injection has been recognized as a viable strategy for CH$_4$ recovery, along with the potential CO$_2$ sequestration (Jiang, 2011; White et al., 2005; Vishal et al., 2015; Chatterjee and Paul, 2013; Mac Dowell et al., 2017; Ampomah et al., 2017; Dai et al., 2016; Dai et al., 2014). A better understanding of CO$_2$/CH$_4$ adsorption is responsible for sustainable shale gas production; and, more importantly, it provides the basic guidelines for optimizing CO$_2$ injection on site for realizing CH$_4$ recovery and CO$_2$ sequestration.

The selective adsorption between CH$_4$ and CO$_2$ is heavily investigated to shed light on the mechanism of shale gas recovery using CO$_2$ injection method. Previously, the CH$_4$ and CO$_2$ adsorption were extensively measured on shale samples; it was found that CO$_2$ presents the adsorption

Confined Fluid Phase Behavior and CO$_2$
Sequestration in Shale Reservoirs
https://doi.org/10.1016/B978-0-323-91660-8.00006-3
187

capacity more than five times greater than CH_4, evaluating the potential of CO_2 for enhanced CH_4 recovery (Nuttall et al., 2005; Kang et al., 2010). Moreover, adsorption of CO_2 and CH_4 was measured on Sichuan basin shale by Duan et al. (2016), and they also observed the much higher adsorption capacity of CO_2 than CH_4. However, CO_2 and CH_4 are generally mingled coming into being gas mixtures, which usually exhibits selective adsorption on shale. Ortiz Cancino et al. (2017) measured the adsorption of an equal molar mixture of CO_2/CH_4 on black shale samples. The measured high adsorption selectivity of CO_2 over CH_4 highlights the more affinity of CO_2 to the organic matter in shale.

Typically, shale rocks mainly comprise organic, such as kerogen, and inorganic matter, such as clay minerals, calcite and quartz. In shale reservoirs, clay minerals in inorganic matter and organic matter are two major controlling factors accounting for gas adsorption (Ji et al., 2012). Lately, the influences of organic substance content, kerogen type and thermal maturity on gas-adsorption has been studied extensively (Zhang et al., 2012). It was found the gas adsorption capacity linearly correlates with the total content of organic matter; meanwhile, high vitrinite accounts for the stronger gas adsorption on shale (Ross and Bustin, 2007a,b; Cheng and Huang, 2004). In addition, clay mineral composition as well as its micropore structure also exerts influences in affecting the gas adsorption owning to their large specific surface area by providing the adsorption sites for natural gas (Cheng and Huang, 2004). Lu et al. (1995) performed adsorption measurements on core samples in Devonian shale; they found that the presence of illite was responsible for adsorption and gas storage, especially for samples with little content of organic matter. In addition, laboratory experiments were performed by Ross and Bustin (2009) to explore the influences of clay minerals, especially montmorillonite and illite, on the storage capacity in shale reservoirs. They deduced that gas adsorption depends on the clay mineral types and the micropore structures of the clay platelets (Ross and Bustin, 2009). Summarily, these studies conducted on shale samples, however, cannot reasonably address the influence of clay minerals on gas adsorption due the coexistence of organic matter. Future studies should be performed to investigate the fundamental mechanisms of the separate effects of typical clay minerals on CO_2/CH_4 adsorption.

Experimental measurements are inevitably sophisticated and generally need complex data-interpretation procedures. By contrast, due to its powerful computational capacity, molecular simulation method is thereby employed as an alternative of the experimental measurement to gain sights into the micro-adsorption behaviors of CO_2/CH_4 in shale systems (Liu and Wilcox, 2012a,b; Psarras et al., 2017; Jin and Firoozabadi, 2014; Huang et al., 2018a,b; Zhang et al., 2015; Kadoura et al., 2016; Yang et al., 2015). Due to presence of organic matter in shale, attempts were employed to explore the selective adsorption of CO_2/CH_4 mixture in organic nanopores using molecular simulations (Huang et al., 2018a,b; Zhang et al., 2015; Kazemi and Takbiri-Borujeni, 2016; Kurniawan et al., 2006; Yuan et al., 2015a,b; Brochard et al., 2012; Wang et al., 2016a,b; Lu et al., 2015; Liu et al., 2016; Sun et al., 2017). Traditionally, graphic and carbon-slit pores are generally used in previous studies to simulate the organic pores in shale (Yuan et al., 2015a,b; Wang et al., 2016a,b). However, these simple carbon-like pores cannot really simulate the natural properties of organic matter in shale. Most recently, studies were conducted to reveal the fundamental mechanisms of CO_2/CH_4 adsorption on realistic organic matter using molecular simulation methods. For example, Sun et al. (2017) used molecular simulations to investigate the adsorption of CO_2/CH_4 in organic nanopores; based on their predicted results, the replacement efficiency of CO_2 was thereby calculated for CH_4 recovery from organic pores. Moisture may play a key role in affecting adsorption of CO_2/CH_4 on shale. Using GCMC simulations, Huang et al. (2018a,b) implemented investigation on the influence of moisture content on the selective adsorption of CO_2/CH_4 on kerogen; a negative effect of the water content was reported to gas adsorption on organic nanopores (Huang et al., 2018a,b). In spite of the previous studies, researches concerning CO_2 injection into clay-rich shales for CH_4 recovery and CO_2 storage are still at a preliminary stage and the underlying mechanisms of CO_2/CH_4 adsorption in clay nanopores are still scarcely reported.

In this section, three typical clay-mineral models, i.e., illite, montmorillonite, kaolinite, are developed using the GCMC simulation method. This method is first validated by comparing the predicted CO_2/CH_4 adsorption with the measured adsorption data. Adsorption behavior of CO_2/CH_4 in clay-mineral nanopores is then predicted under different reservoir conditions; specifically, the influencing factors, i.e., system pressure, temperature, and pore size, are comprehensively discussed from the micropore-scale perspective. Furthermore, we have a deep discussion on the potential implications of CO_2 injection for CH_4 recovery and CO_2 sequestration by calculating the adsorption selectivity of CO_2/CH_4 in the clay-mineral pores. This study is useful for addressing the following questions: (1) can molecular simulation reasonably represent the adsorption behavior of gas mixtures on clay minerals? (2) how does the CO_2/CH_4 adsorption behave in different clay-mineral pores? (3) how to reasonably optimize CO_2 injection to achieve successful CO_2 sequestration and CH_4 recovery in clay-rich shales with the knowledge of CO_2/CH_4 adsorption on clay-minerals? This study helps gain a deep insight into the adsorption behavior of CO_2/CH_4 in clay-rich gas-bearing reservoirs; more importantly, it provides the guidelines for future optimization design of CO_2 injection for CH_4 recovery and CO_2 sequestration in the field applications.

5.1.1 Molecular clay-mineral models

Based on the X-ray diffraction (XRD) analyses, illite, montmorillonite, and kaolinite are three dominating clay minerals in Longmaxi shale samples derived from Sichuan Basin of China (Wang et al., 2016a,b; Lu et al., 2015). According to surface morphology of the Longmaxi shale, shape morphology of the mineral pores in illite, montmorillonite and kaolinite is simplified as slit-shaped.

Illite is one typical 2:1 clay mineral in shale, which is consisted of one Al-O octahedral sheet located between two Si-O tetrahedral sheets (Deer et al., 1996). The illite is represented with the general unit cell formula of $K_x[Al_xSi_{(8-x)}][Al_yMg_{4-y}]O_{20}(OH)_4$ (Szczerba et al., 2015), where x and y are set as 1 and 4, respectively, for each unit cell. In every unite cell, one Si^{4+} is only replaced by only one Al^{3+}, resulting in negative charged clay sheets. K^+ cations are randomly distributed between the interlayers, counterbalancing the induced electrostatic charges in the unit cell. It is noted that K^+ cations are mobile in the simulation cell. The parameters of each unit cell are selected as $a=0.516$ nm, $b=0.935$ nm, $\alpha=91.03°$, $\beta=100.37°$, $\gamma=89.75°$ (Sainz-Diaz et al., 2003; Wardle and Brindley, 1972). The simulation cell contains two clay sheets, wherein each sheet consists of 24 unit cells ($6 \times 4 \times 1$ supercell). The simulation cell thereby has the dimension of 3.096 nm $\times 3.740$ nm in the x and y directions, respectively. The illite nanopores are represented by two clay sheets separated with a fixed distance. Fig. 5.1A presents the schematic structure of the illite nanopore. Pore width is calculated as the separation between the centers of two opposite oxygen atoms in the inner layers of the two clay sheets.

Montmorillonite is a 2:1 cationic mineral in shale, which is comprised by two Si-O sheets and one Al-O sheet (Wang et al., 2016a,b). The simulation cell contains two clay sheets, wherein each sheet consists of 24 unit cells ($6 \times 4 \times 1$ supercell). The simulation cell has the dimension

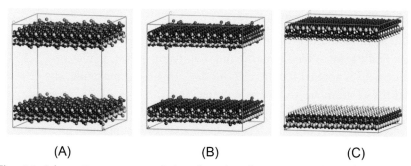

Fig. 5.1 Schematic structures of the clay-mineral nanopores: (A) K-illite; (B) Na-montmorillonite; and (C) kaolinite. *Red* spheres represent O atoms, *pink* spheres represent Al atoms, *green* spheres represent Mg atoms, *yellow* spheres represent Si atoms, white spheres represent H atoms, *blue* spheres represent K$^+$ ions, and *cyan* spheres represent Na$^+$ ions.

of 3.168 nm × 3.656 nm in the x and y directions, respectively. Within the supercell, one Al^{3+} atom in every eight Al^{3+} atoms is homogeneously substituted by Mg^{2+} in the octahedral sheet; moreover, one Si^{4+} atom in every 32 atoms is replaced by Al^{3+} atoms in the tetrahedral sheet (Chávez-Páez et al., 2001), resulting in negative charged clay sheets. The negative charge is counterbalanced by introducing Na$^+$ cations which are randomly distributed between the interlayers, resulting in Na-montmorillonite with the formula of Na$_{0.75}$(Al$_{0.25}$Si$_{7.75}$)(Al$_{3.5}$Mg$_{0.5}$)O$_{20}$(OH)$_4$. The montmorillonite nanopores are represented by two clay sheets separated with a fixed distance, of which the pore width is calculated as the separation between the centers of two opposite oxygen atoms in the inner layers of the two clay sheets. Fig. 5.1B presents the schematic structure of the montmorillonite nanopore.

Kaolinite is a 1:1 clay mineral, which is consisted of one Al–O octahedral sheet and one Si–O tetrahedral sheet. The parameters of the unite cell are set as $a = 0.5154$ nm, $b = 0.8942$ nm, $c = 0.7391$ nm, $\alpha = 91.92°$, $\beta = 105.05°$, $\gamma = 89.90°$ (Tenney and Cygan, 2014), resulting in the formula of Al$_4$Si$_4$O$_{10}$(OH)$_8$ (Tenney and Cygan, 2014). Each sheet consists of 24 unit cells (6 × 4 × 1 supercell), resulting a dimension of 3.0924 nm × 3.5768 nm in the x and y directions, respectively. The kaolinite nanopores are represented by two clay sheets separated with a fixed distance, of which the pore width is calculated as the separation between the centers of two opposite hydrogen atoms in the two clay sheets. Fig. 5.1C presents the schematic structure of the kaolinite nanopore. It is noted that the equilibrium occurs at a point where the adsorption and desorption have the same grand potential values.

5.1.2 Force field parameters

The potential models used for CH_4 and CO_2 molecules are obtained from the TraPPE force filed (Martin and Siepmann, 1998), while the CLAYFF force field is applied for the clay models. In a CH_4 molecule, the C atom and the H atom are treated as a united atom. As for a CO_2 molecule, the partial charge of the C atom and O atom are $+0.7\,e$ and $-0.35\,e$, respectively. The CLAYFF model has been extensively applied to investigate the thermodynamic as well as the static properties of fluids in clay minerals, which presents an excellent agreement with the experimental data (Cygan et al., 2004). The nonbonded interactions include Lennard-Jones (LJ) and electrostatic terms, which are described by the following potential model,

$$u\left(r_{ij}\right) = 4\varepsilon_{ij}\left[\left(\frac{\sigma_{ij}}{r_{ij}}\right)^{12} - \left(\frac{\sigma_{ij}}{r_{ij}}\right)^{6}\right] + \frac{q_i q_j}{4\pi\varepsilon_0 r_{ij}} \tag{5.1}$$

where r_{ij} represents the separation distance between the atoms i and j; ε_{ij} and σ_{ij} represent the well depth of LJ potential and the LJ radius, respectively; q is the atrial atom charge to calculate the Coulomb interactions. The Lorentz-Berthelot combining rules are used to calculate the cross interactions between two different atoms (Lorentz, 1881),

$$\sigma_{ij} = \frac{1}{2}\left(\sigma_{ii} + \sigma_{jj}\right) \tag{5.2}$$

$$\varepsilon_{ij} = \sqrt{\varepsilon_{ii}\varepsilon_{jj}} \tag{5.3}$$

The cutoff radius is set as $1.25\,nm$. In order to take the long-range electrostatic interactions into consideration, a vacuum slab is placed in the z direction in the simulation cell, which is estimated with the Ewald summation method (Jin and Firoozabadi, 2014; Crozier et al., 2000).

5.1.3 Simulation details

GCMC simulations are applied to explore adsorption behavior of the CO_2/CH_4 mixture in the clay-mineral nanopores under shale reservoir conditions. This work is performed in the grand canonical μVT ensemble; it is noted that within such an ensemble, the chemical potentials (μ), system volume (V) and temperature (T) are kept as constant (Liu et al., 2018a,b). Within the framework of the GCMC simulations, gas molecules confined in the clay-mineral pores can exchange with a fictitious bulk gas reservoir; such a reservoir is set with a fixed chemical potential. Notes that the chemical potentials are computed using the Peng-Robinson equation of state (Lu et al., 2015). An equal probability is endowed to the gas molecules, which can be inserted or removed from the confined nanopores depending on the chemical potential of the fictitious gas reservoir. This simulation comprises of 0.2 million cycles for each molecule to achieve equilibrium and 0.7 million cycles for sampling fluid distribution in nanopores.

The excess adsorption is computed by deducting the total amount of gas, which occupies the entire pore volume and is endowed with the bulk density, from the total gas loading (Huang et al., 2018a,b; Jin and Firoozabadi, 2014),

$$\Gamma_{ex} = \frac{\langle N_{ab} \rangle}{N_A} - \frac{\rho_{bulk} V}{M} \qquad (5.4)$$

where Γ_{ex} represents the amount of excess adsorption, kmol/m³; $\langle N_{ab} \rangle$ represents the total amount of gas residing in clay-mineral pores; N_A represents Avogadro constant, 6.022×10^{23}; M represents molecular weight, g/mol. V represents the effective pore volume. ρ_{bulk} denotes the bulk gas density according to the National Institute of Standards and Technology (NIST) Chemistry WebBook, g/cm³.

The excess adsorption calculation highly depends on the determination of the effective pore volume. Based on the volumetric method, the effective pore volume is determined from the helium adsorption by assuming that helium adsorption is negligible in nanopores and the total uptake is dominated by the mechanism of pore filling (Tian et al., 2017). It has been reported that the pore volume of different clay-mineral pores can remain in a constant value in a wide temperature and pressure range (Zhang et al., 2012); the effective pore volume is then calculated using the helium adsorption as,

$$V_p = \frac{\langle N_{He} \rangle}{N_A \rho_{He, bulk}^{molar}} \qquad (5.5)$$

where $\langle N_{He} \rangle$ is the average molecular numbers of helium in nanopores; $\rho_{He, bulk}^{molar}$ represents the molar density of helium in bulk; and N_A represents Avogadro constant. It is found that the effective pore volume is independent on the system pressure and temperature, while it increases linearly with slit aperture. In addition, the effective pore volume is less than that obtained by simply multiplying the surface area with the slit aperture due to the finite molecular size of helium (Tian et al., 2017).

The adsorption selectivity of CO_2 over CH_4 is calculated to characterize the selective adsorption behavior of CO_2/CH_4 mixture in clay nanopores, which is calculated as (Kurniawan and Bhatia, 2006),

$$S_{CO_2/CH_4} = \frac{(x_{CO_2}/x_{CH_4})}{(y_{CO_2}/y_{CH_4})} \tag{5.6}$$

where x_{CO_2} and x_{CH_4} represent the molar fractions of CO_2 and CH_4 in the clay nanopores, respectively; y_{CO_2} and y_{CH_4} represent the molar fractions of CO_2 and CH_4 in the bulk gas reservoir. If $S_{CO_2/CH4}$ is larger than unit, it suggests that CO_2 is more inclined to adsorb on the clay surface than CH_4 (Liu and Wilcox, 2012a,b).

5.1.4 Validation of the GCMC method

The excess adsorption of CO_2/CH_4 mixture (50.0 mol%:50.0 mol%) are measured on the three clay minerals using the thermogravimetric method. The thermogravimetric method can be inferred in our previous works (Liu et al., 2018a,b). In addition, the three clay minerals are montmorillonite, illite and kaolinite, which are pure powder samples. The measured data is then compared with that predicted results from the GCMC simulations to validate its efficiency in calculating CO_2/CH_4 adsorption on clay minerals, as shown in Figs. 5.2 and 5.3. It should be noted that the CO_2/CH_4 mixture with the same composition is built in the GCMC simulation; meanwhile,

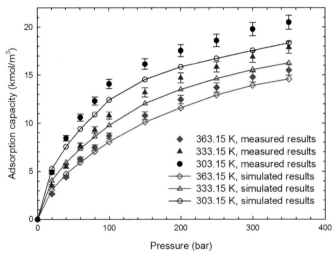

Fig. 5.2 Adsorption capacities of the CO_2/CH_4 (50.0 mol%:50.0 mol%) mixture on montmorillonite.

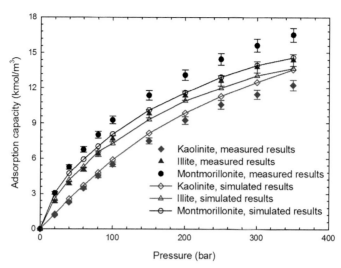

Fig. 5.3 Adsorption capacities of CO_2/CH_4 (50.0 mol%:50.0 mol%) mixture on montmorillonite, illite and kaolinite at 333.15 K.

the adsorption of CO_2/CH_4 mixture is simulated in the three clay mineral pores to mimic the experimental measurement.

The calculated excess adsorption of CO_2/CH_4 mixture on the three clay minerals does not fit perfectly well with the measured adsorption data, especially as higher-pressure conditions. However, the magnitude of excess adsorption of CO_2/CH_4 mixture is fairly close to the experimental results. As shown in Fig. 5.2, the deviation between the experimental and simulation results is larger at 303.15 K, indicating that the GCMC simulation method exhibits less efficiency at low temperature conditions. Fig. 5.3 presents that the calculated excess adsorption of CO_2/CH_4 mixture show the highest capacity on montmorillonite, followed by illite and kaolinite, at the same temperature/pressure conditions, which is consistent with the measured adsorption isotherms. The difference between the measured and simulated adsorption is probably resulted from the issue of pore accessibility (Bae et al., 2009, 2014; Liu et al., 2018a,b; Nguyen and Bhatia, 2007, 2008; Bae and Bhatia, 2006); specifically, due to the presence of impenetrable pore necks between theoretically enterable pores, adsorbates generally have the pore accessibility problems into the realistic heterogeneous porous materials, such as shale and clay samples (Huang et al., 2018a,b). Based on the statistical analysis, the relative deviations between the calculated and the experimental results are within an acceptable range of 0%–5.25%, validating our GCMC simulations in predicting excess adsorption of CO_2/CH_4 mixture under shale reservoir conditions.

5.1.5 CH_4/CO_2 adsorption on clay minerals

5.1.5.1 Effect of system pressure

Figs. 5.4–5.6 present the density profiles of CO_2/CH_4 (50.0 mol%: 50.0 mol%) in the 3-nm clay-mineral pores at different pressures. In the first place, the density near the pore surface formed by CO_2 is significantly higher

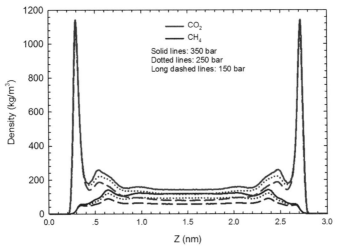

Fig. 5.4 Density profiles of CO_2/CH_4 (50.0 mol%:50.0 mol%) at 363.15 K in the 3-nm illite pore.

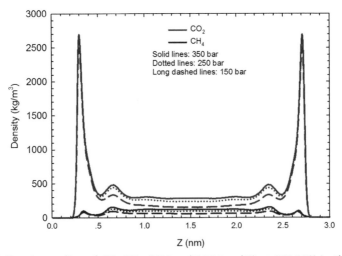

Fig. 5.5 Density profiles of CO_2/CH_4 (50.0 mol%:50.0 mol%) at 363.15 K in the 3-nm montmorillonite pore.

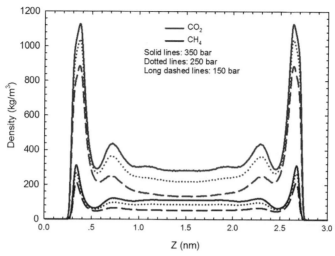

Fig. 5.6 Density profiles of CO_2/CH_4 (50.0 mol%:50.0 mol%) at 363.15 K in the 3-nm kaolinite pore.

than that by CH_4, suggesting the much stronger adsorption capacity and the more affinity of CO_2 in clay nanopores. More interestingly, CH_4 performs the highest adsorption capacity on kaolinite, accompanying with the formation of stronger adsorption layers on pore surface. On the contrary, the adsorption capacity of CO_2 on clay minerals are calculated in the order of montmorillonite > illite ≈ kaolinite. That is, montmorillonite has the highest adsorption to CO_2 but the smallest adsorption capacity to CH_4. It implies that as for montmorillonite-rich shale, CO_2 injection could be a more efficient approach for CH_4 recovery compared to illite- and kaolinite-rich shale reservoirs. Kaolinite consists of one Al-O octahedral sheet and one Si-O tetrahedral sheet; CH_4 molecules tend to be more attracted on kaolinite due to the stronger fluid–fluid interactions compared to that on illite and montmorillonite. However, the interaction between kaolinite and CO_2 molecules is weaker than that on illite and montmorillonite, resulting in the less adsorption on kaolinite.

Moreover, as shown in Figs. 5.4–5.6, density of CO_2 near the pore surface is reduced with the decreasing pressure in the kaolinite nanopores, suggesting that CO_2 adsorption on kaolinite is sensitive to reservoir pressures; by contrast, the CO_2 density near the pore surface is slightly affected by the pressure change in the illite and montmorillonite nanopores, indicating the less sensitivity of CO_2 to pressure on illite and montmorillonite. In addition, CH_4 adsorption on clay minerals is closely correlated with the reservoir pressure, indicated by the decrement of CH_4 density near the pore surface with the decreasing pressure. From this view, it is inferred that CO_2 injection could be most efficient for CH_4 recovery in the montmorillonite-rich shale gas reservoir, while it is the least efficient for kaolinite-rich shale reservoirs.

5.1.5.2 Effect of system temperature

Figs. 5.7–5.9 present the density profiles of CO_2/CH_4 (50.0 mol%: 50.0 mol%) in the 3-nm clay-mineral pores at 300 bar. Besides the adsorption layer formed near the pore surface, a second adsorption layer is also resulted from CO_2 adsorption in the clay-mineral nanopores. As system temperature decreases, the averaged density of CO_2 and CH_4 is reduced in the clay-mineral pores, indicating that adsorption of CO_2 and CH_4 is affected by the reservoir temperature. During shale reservoir development, reservoir temperature is generally kept as constant; the influence of temperature on shale gas production is thus rarely considered in previous works (Liu and Hou, 2019). However, adsorption/desorption of shale gas during exploration is a dynamic process, accompanying with heat produce and release, which exerts a big impact on the reservoir temperature. It is thereby essential for revealing the effect of temperature on the adsorption behavior of CO_2/CH_4 mixture on shale.

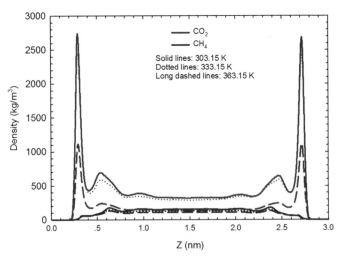

Fig. 5.7 Density profiles of CO_2/CH_4 (50.0 mol%:50.0 mol%) in the 3-nm illite pore at 300 bar.

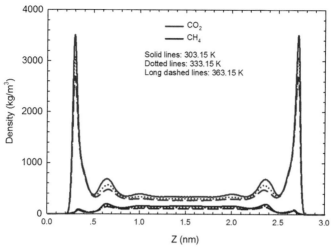

Fig. 5.8 Density profiles of CO_2/CH_4 (50.0 mol%:50.0 mol%) in the 3-nm montmorillonite pore at 300 bar.

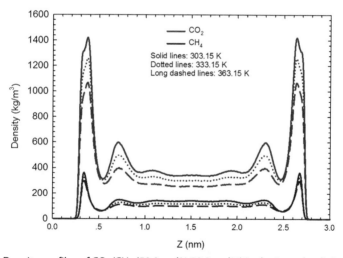

Fig. 5.9 Density profiles of CO_2/CH_4 (50.0 mol%:50.0 mol%) in the 3-nm kaolinite pore at 300 bar.

5.1.5.3 Effect of pore size

One unique characteristic of shale reservoirs is heterogeneity, where pores in shale matrix exhibit pore-size distributions. Previously, extensive studies were conducted to reveal the adsorption behavior of shale gas in single nanopores. The adsorption behavior of CO_2/CH_4 mixture is investigated in clay-mineral pores with different pore sizes. Figs. 5.10–5.12 present

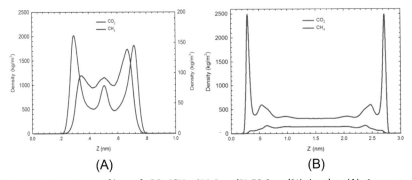

Fig. 5.10 Density profiles of CO_2/CH_4 (50.0 mol%:50.0 mol%) in the (A) 1-nm and (B) 3-nm illite pores at 333.15 K and 350 bar.

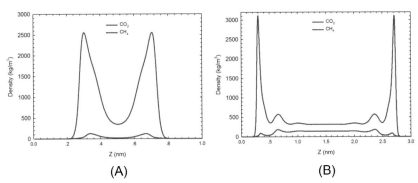

Fig. 5.11 Density profiles of CO_2/CH_4 (50.0 mol%:50.0 mol%) in the (A) 1-nm and (B) 3-nm montmorillonite pores at 333.15 K and 350 bar.

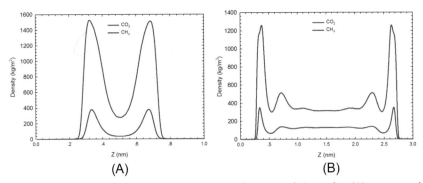

Fig. 5.12 Density profiles of CO_2/CH_4 (50.0 mol%:50.0 mol%) in the (A) 1-nm and (B) 3-nm kaolinite pores at 333.15 K and 350 bar.

the density profiles of CO_2/CH_4 (50 mol%:50 mol%) mixture in 1- and 3-nm pores, which are comprised of different clay minerals, i.e., illite, montmorillonite, and kaolinite, respectively. The maximum adsorption density of CH_4 in the 1-nm pores is almost the same as that in the 3-nm pores. However, the maximum adsorption density of CO_2 in the 3-nm pores, comprised of montmorillonite and kaolinite, is remarkably higher than that in the 1-nm pores. From this view, CO_2 is superior in the large pores to that in the small pores for CH_4 recovery.

As shown in Figs. 5.10–5.12, density in the center of 1-nm pore is generally higher than that of the 3-nm pore due to the coupling effect from both sides of pore walls. In addition, a second adsorption layer comes into being in the 1-nm illite pore by CO_2 and CH_4 molecules. As shown in Fig. 5.10, quite interestingly, the adsorption layers formed by CO_2 and CH_4 are not equally distributed in both sides of the 1-nm illite pore; moreover, the formed adsorption layers by CO_2 are complementary to that by CH_4 molecules. On the contrary, such asymmetric adsorption behavior is not observed in montmorillonite and kaolinite pores. This unique adsorption behavior occurred in illite is possibly caused by the preferential adsorption between CH_4 and CO_2 molecules on the negative charged pore surfaces when pore size is lower than certain value. In addition, the adsorption layers formed in the 1-nm pores are thicker than those formed in the 3-nm pores. In other words, the adsorbed state gas tends to dominate in the smaller clay nanopores, which are thereby the ideal space for CO_2 sequestration. As for clay-rich shale gas reservoirs, we should pay heavy attention to how to efficiently recover the adsorbed gas reserves from the smaller pores.

5.1.6 Implications for CH_4 recovery and CO_2 sequestration

Adsorption of CO_2/CH_4 mixture in clay-mineral pores closely correlates the implications of CO_2 injection for CH_4 recovery and CO_2 sequestration under shale reservoir conditions. In this section, the results obtained from the previous sections are used to analyze the implication potentials of CH_4 recovery and CO_2 sequestration in clay-rich shale reservoirs. Figs. 5.13–5.15 present the adsorption selectivity of CO_2/CH_4 (50.0 mol%:50.0 mol%) in the 3-nm clay-mineral pores. It is reported that the most important criteria for CO_2 sequestration is whether the buried CO_2 can realize the long-time storage underground or not (Bachu, 2002). As shown in Figs. 5.13–5.15, illite and montmorillonite present higher adsorption selectivity than kaolinite at the same pressure and temperature conditions; it implies that the adsorption capacity of CO_2 relative to CH_4 is stronger on illite and montmorillonite than that on kaolinite. Such a high adsorption capacity of CO_2 on the clay

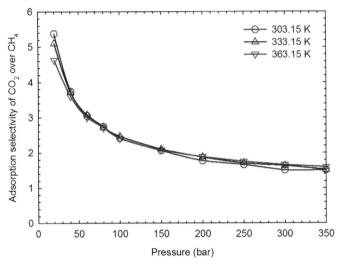

Fig. 5.13 Adsorption selectivity of CO_2/CH_4 (50.0 mol%:50.0 mol%) in the 3-nm illite pore.

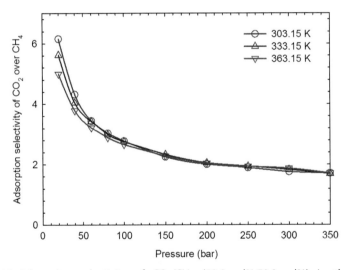

Fig. 5.14 Adsorption selectivity of CO_2/CH_4 (50.0 mol%:50.0 mol%) in the 3-nm montmorillonite pore.

minerals facilitates CO_2 sequestration for a long-time storage in shale gas reservoir, while the relatively higher adsorption of CO_2 to CH_4 enables CO_2 a much high efficiency for CH_4 recovery. Illite and montmorillonite are thereby adopted as the optimized clay mineral types for CO_2 sequestration and CH_4 recovery in the field applications.

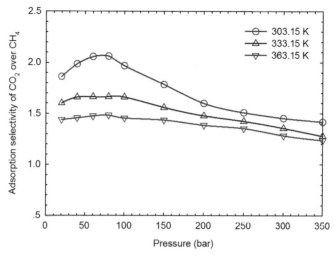

Fig. 5.15 Adsorption selectivity of CO$_2$/CH$_4$ (50.0 mol%:50.0 mol%) in the 3-nm kaolinite pore.

Understanding of the effect of pressure on selective adsorption of CO$_2$ over CH$_4$ can provide the basic guidance for CO$_2$ injection with the implications of CO$_2$ sequestration and CH$_4$ recovery. Results show that the selective adsorption of CO$_2$ over CH$_4$ decreases in the illite and montmorillonite pores as reservoir pressure increases. However, as for kaolinite pores, the adsorption selectivity is generally larger at lower pressure conditions, and there is an optimum pressure (around 70 bar) for the maximum adsorption selectivity. Previous works also observed the optimum pressure on other materials for the adsorption selectivity of CH$_4$ and CO$_2$ (Vieira et al., 2018). The different adsorption selectivity of CO$_2$ over CH$_4$ on kaolinite results from the relatively higher adsorption capacity of CO$_2$ over CH$_4$ on the kaolinite mineral at the medium pressure conditions. During shale gas reservoir development, reservoir pressure decreases when CH$_4$ is produced from shale reservoirs. However, a large amount of adsorbed CH$_4$ is still left in the depleted shale gas reservoirs if no further exploitation was implemented. Based on the results, the injection pressure of CO$_2$ should be around 60 bar for kaolinite-rich shale, while the pressure for illite and montmorillonite-rich shales are around 30 bar, to realize the highest efficiency for CH$_4$ recovery and CO$_2$ sequestration in field applications.

Figs. 5.16–5.18 present the adsorption isotherms of CO$_2$/CH$_4$ (50.0 mol%: 50.0 mol%) mixture and the individual adsorption of CH$_4$ and CO$_2$ in the

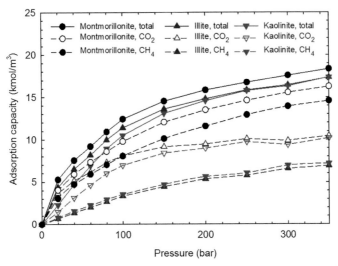

Fig. 5.16 Adsorption isotherms of CO_2/CH_4 (50.0 mol%:50.0 mol%) mixture and the individual adsorption of CH_4 and CO_2 in the 3-nm mineral pores at 303.15 K.

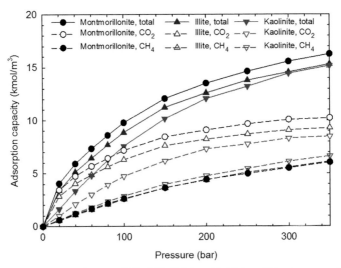

Fig. 5.17 Adsorption isotherms of CO_2/CH_4 (50.0 mol%:50.0 mol%) mixture and the individual adsorption of CH_4 and CO_2 in the 3-nm mineral pores at 333.15 K.

3-nm mineral pores. CO_2/CH_4 mixture has the adsorption capacity in the clay-mineral pores in the order of montmorillonite > illite > kaolinite. When pressure is low, illite exhibits stronger adsorption capacity to gas mixtures than kaolinite; however, as system pressure increases, adsorption capacity of the gas mixture on kaolinite approaches that on illite. Moreover, CO_2 has the highest

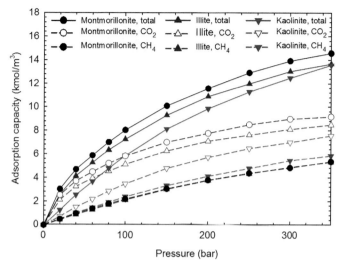

Fig. 5.18 Adsorption isotherms of CO_2/CH_4 (50.0 mol%:50.0 mol%) mixture and the individual adsorption of CH_4 and CO_2 in the 3-nm mineral pores at 363.15 K.

adsorption capacity on the montmorillonite pore surface; on the contrary, montmorillonite shows the smallest adsorption capacity for CH_4. It validates the point that the montmorillonite-rich shale is the most ideal reservoir for CO_2 sequestration and CO_2 injection can achieve the best recovery efficiency of CH_4 in montmorillonite-rich shale reservoirs. By contrast, kaolinite has the smallest individual adsorption for CO_2 but the highest adsorption for CH_4, indicating that kaolinite-rich shales is unfavorable for CH_4 recovery and CO_2 sequestration.

Collectively, this work provides important insights into the mechanism of selective adsorption of CH_4 and CO_2 on tight clay-rich shale, which is beneficial to shale gas reservoir assessment and shale gas production optimization using CO_2 injection method. It is significantly economical for shale gas recovery in clay-rich shale reservoirs by optimizing CO_2 injection with the knowledge of the adsorption behavior of CH_4 and CO_2 on clay nanopores. In addition to CH_4, some heavier components, such as C_2H_6, C_3H_8, etc., may also be dominating in shale gas. Thereby, future works are recommended to investigate the selective adsorption behavior between CO_2 and the heavier hydrocarbons. In addition, the adsorption sites may also play a key role in affecting the adsorption behavior of CO_2 and CH_4 on shale. In future works, this topic should also be investigated.

5.2 Comparing the effectiveness of SO_2 with CO_2 for replacing hydrocarbons from nanopores

Carbon dioxide (CO_2) and sulfur dioxide (SO_2) in flue gas are the main components causing air pollutions. CO_2 that causes global warming is attracting extensive attention, while SO_2 that may cause acid rain and ozone depletion is strictly restricted for emission (Yi et al., 2014). Recently, many studies have been conducted to remove CO_2 and SO_2 from flue gas independently (Yi et al., 2014; Lopez et al., 2007; Liu et al., 2009; Lee et al., 2002; Ziołek et al., 2000; Juray and Guy, 2000; Bagreev et al., 2002; Sumathi et al., 2010a,b; Siriwardane et al., 2001). However, the methods proposed in these studies have the disadvantages of either high-energy consumption or expensive equipment cost. The injection of flue gas into shale reservoirs may be a promising way in removing CO_2 and SO_2 by realizing CO_2 and SO_2 sequestration underground; on the other hand, shale hydrocarbons can be replaced due to the competitive adsorption between hydrocarbons, CO_2 and SO_2 on shale.

Shale hydrocarbon is one kind of important unconventional resources; due to its considerable abundance, studies regarding to recover such resources from shale reservoirs have been extensively conducted (Huang et al., 2018a,b; Yuan et al., 2015a,b; Weijermars, 2014; Yamazaki et al., 2006; Karacan et al., 2011). However, the unique characteristics of shale reservoirs, such as extremely low permeability, and heterogeneity, make it difficult to recover shale resources from such reservoirs (Weijermars, 2013). Unlike conventional reservoirs, pore size in shale matrix is generally in the nanoscale; fluids confined in such pores are strongly attracted by the pore surface. Moreover, shale may also contain a large proportion of organic matter, such as kerogen, while hydrocarbons have the affinity to the organic matter. Due to the existence of organic matter as well as nanopores, hydrocarbons tend to adsorb on the organic pore surface, showing unique adsorption behavior from that in conventional reservoirs due to the stronger fluid-pore surface interactions. A better understanding of the adsorption behavior of fluid in organic pores is beneficial in obtaining the fundamental mechanisms of shale-hydrocarbon storage in shale reservoirs.

The idea of CO_2 injection into shale reservoirs has been proposed as a feasible technique to enhance shale hydrocarbon recovery; extensive studies have been conducted on this subject, although this technique has not been widely commercialized. Recently, studies regarding the enhanced shale

C_1 recovery using CO_2 method are mostly conducted considering that C_1 is generally the most commonly seen component in shale gas. Numerical simulations were performed, and it has been found that introduction of CO_2 into the depleted shale gas reservoirs is technically feasible for enhancing shale C_1 recovery (Kim et al., 2017; Luo et al., 2013; Yu et al., 2014; Jiang et al., 2014; Godec et al., 2013; Liu et al., 2013). To understand the fundamental mechanisms of recovering C_1 using CO_2 injection from organic shale, experimental studies were conducted on the adsorption behavior of C_1/CO_2 on typical shale samples (Gensterblum et al., 2014; Ottiger et al., 2008; Bhowmik and Dutta, 2011; Faiz et al., 2007; Khosrokhavar et al., 2014; Busch et al., 2003; Majewska et al., 2009; Ross and Bustin, 2007a,b). It has been found that CO_2 exhibits a higher adsorption capacity than C_1, suggesting that CO_2 could be an effective agent to recover shale C_1 from shale reservoirs (Gensterblum et al., 2014; Ottiger et al., 2008; Bhowmik and Dutta, 2011; Faiz et al., 2007; Khosrokhavar et al., 2014; Busch et al., 2003; Majewska et al., 2009; Ross and Bustin, 2007a,b). Compared with the experimental measurements, molecular simulation is a powerful theoretical approach to gain insights into the microphase behavior of gas mixtures from the molecular perspective. Using molecular simulation, many attempts have been conducted to investigate the competitive adsorption behavior of C_1/CO_2 mixtures in organic pores (Zhang et al., 2015; Kazemi and Takbiri-Borujeni, 2016; Kurniawan et al., 2006; Yuan et al., 2015a,b; Brochard et al., 2012; Wang et al., 2016a,b; Lu et al., 2015; Liu et al., 2016; Sun et al., 2017; Huang et al., 2018a,b; Wu et al., 2015a, b; Kowalczyk et al., 2012). Based on the simulation results, CO_2 presents stronger adsorption on pore surface than that of C_1 (Ambrose et al., 2012).

Besides C_1, heavier hydrocarbons, such as C_2, nC_3, nC_4, and nC_5, generally coexist, which usually show individual adsorption capacities on organic shale. Jin and Firoozabadi (2014) studied the influence of CO_2 on the adsorption of different hydrocarbons (i.e., C_1 and nC_4) in nanopores. Compared to C_1, adsorbed nC_4 cannot be readily replaced by CO_2 due to the stronger associations of nC_4 to the organic pore surface. To date, studies regarding the recovery of heavier hydrocarbons from shale reservoirs using CO_2 are still at an initial stage. Recently, experimental studies were conducted to compare the adsorption of SO_2 and CO_2 on some commercial adsorbents (Dong et al., 2017; Luo et al., 2017). It was found that adsorption of SO_2 on the surface of commercial adsorbents is significantly higher than that CO_2. It inspires us that SO_2 may have a distinguished performance in replacing hydrocarbons from organic shale. To the best of our knowledge, quite limited studies are conducted to investigate the efficiency of SO_2 for enhanced shale hydrocarbon recovery.

Based on the previous study, SO_2 has been experimental found to be a more efficient agent than CO_2 for recovering C_1 and C_2 from shale samples using the low-field nuclear magnetic resonance technique (Huang et al., 2019a,b). In this section, CO_2 and SO_2 is introduced into a "hydrocarbon-saturated" (i.e., saturated with C_1, C_2, nC_3, nC_4, and nC_5) 3-nm pore to investigate how CO_2 and SO_2 affect the fluid distribution of hydrocarbons in nanopore. Based on the altered fluid distribution, the selectivity of CO_2 and SO_2 over different hydrocarbons is then calculated to analyze the competitive adsorption of CO_2, SO_2, and hydrocarbons in their binary mixtures in the organic pore. The replacement efficiency of CO_2 and SO_2 over these hydrocarbons is then calculated to compare the usage of CO_2 and SO_2 for enhanced shale hydrocarbon recovery.

The objective of this section is to propose SO_2 injection as a new strategy for enhancing shale hydrocarbon recovery and to assessing the efficiency of CO_2 for the recovery of heavier hydrocarbons from organic shale. Carbon slit-pore has been widely used in the simulation works to represent kerogen walls in shale, considering that carbon surface is hydrocarbon–wet and can provide underlying mechanisms on the adsorption behavior of CH_4 in nanopores (Ambrose et al., 2012). As part of a comprehensive study on the adsorption behavior of gas on organic shale, we also use a carbon slit-pore model to describe nanopores for simplicity. This work is expected to provide the basic understanding of competitive adsorption behaviors of hydrocarbons, CO_2 and SO_2 in organic pores, wherein it helps to evaluate the efficiency of CO_2 and SO_2 in replacing hydrocarbons from organic pores and thereupon realize CO_2 and SO_2 sequestration.

5.2.1 Molecular dynamics

The Forcite Module and Amorphous Cell Package is applied to conduct the MD simulations (Yu et al., 2017; Zhao et al., 2016; Valentini et al., 2011; Li et al., 2016). In the Force Module, the condensed–phased–optimized molecular potential for atomistic simulation studies (COMPASS) force field is employed to describe the interatomic interactions; such force field has been recognized as the first force-field that enables an accurate simultaneous prediction for a wide range of molecules (Li et al., 2016; Rigby et al., 1997). In the COMPASS force field, the total potential energy (E^{total}) is expressed as (Rigby et al., 1997):

$$E^{total} = E^{internal} + E^{cross-coupling} + E^{vanderWaals} + E^{electrostatic} \quad (5.7)$$

$$E^{internal} = \sum E^{(b)} + \sum E^{(\theta)} + \sum E^{(\phi)} + \sum E^{(\gamma)} \quad (5.8)$$

$$E^{cross-coupling} = \sum E^{(b\theta)} + \sum E^{(b\phi)} + \sum E^{(b'\phi)} + \sum E^{(\theta\theta')}$$
$$+ \sum E^{(\theta\phi)} + \sum E^{(\theta\theta'\phi)} \quad (5.9)$$

where b and b' are the lengths of two adjacent bonds; θ and θ' are the angles between two adjacent bonds; ϕ is the angle resulted from dihedral torsion; and γ is the out of the plane angle (Li et al., 2016). $E^{internal}$ is the energy derived from each of the internal valence coordinates; $E^{cross-coupling}$ is the cross-coupling term between internal coordinates. $E^{vanderWaals}$ is the sum of repulsive and attractive Lennard-Jones terms (Jones, 1924). The $E^{vanderWaals}$ and $E^{electrostatic}$ are obtained according to the atom-based method with a cutoff distance of 13.0 Å (Song et al., 2017). In addition, we use the Andersen thermostat (Andersen, 1980) for the temperature conversion.

5.2.2 Simulation model

Carbon materials are generally applied to simulate organic matter in shale considering that carbon surface is hydrocarbon-wet and can provide the underlying mechanisms of adsorption of hydrocarbons in organic pores (Ambrose et al., 2012; Li et al., 2014; Liu et al., 2018a,b). In this simulation, the full atomistic structure of graphite layers, formed by carbon atoms, is used to simulate the organic nanopore. As shown in Fig. 5.19, one carbon-slit pore is connected by a fictitious reservoir with a given volume and temperature. CO₂ or SO₂ is firstly introduced into the fictitious reservoir, which can spontaneously enter the organic nanopore due to the affinity

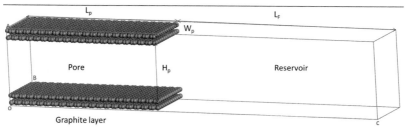

Fig. 5.19 Schematic of the 3-nm nanopore in the MD simulation.

to the pore surface. The periodic boundary condition is applied for the created pore in all three directions. Moreover, two graphic layers are used to form one carbon sheet. The separation distance between two carbon–atom centers in two opposite graphite layers is 0.335 nm. In the same graphite layer, the distance between two adjacent carbon–atom centers is 0.142 nm. During the MD simulations, the position of each carbon sheet is fixed. The size of the created pore is $(L_p + L_F)$ nm $\times W_P$ nm $\times (H_P + 2W_c)$ nm in the OC, OB, and OA directions, respectively (L_p is the length of the nanopore, 6.619 nm; L_F is the length of the fictitious reservoir; W_p is the width of the nanopore, 2.952 nm; and W_c is the separation distance between the two carbon–atom centers in the two graphite layers, 0.335 nm).

MD simulations are performed in a canonical NVT ensemble, which has fixed number of particles (N), volume (V), and temperature (T). A binary mixture composed of given number of molecules is initially loaded in the NVT ensemble. The dynamic density distribution of each species is obtained after achieving competitive adsorption/desorption equilibrium of the binary mixture on the pore surface. The basic inputs in the MD simulations are summarized in Table 5.1.

When placed in this nanopore, C_1, C_2, nC_3, nC_4, or nC_5, CO_2, and SO_2 molecules in their binary mixtures will exhibit competitive adsorption on the organic pore surface. In this work, adsorption selectivity is used to characterize the competitive adsorption of C_1, C_2, nC_3, nC_4, nC_5, CO_2, and SO_2 molecules on the pore surface. As for binary mixtures, the adsorption selectivity ($S_{A/B}$) is calculated as (Kurniawan et al., 2006),

$$S_{A/B} = \frac{(x_A/x_B)}{(y_A/y_B)} \qquad (5.10)$$

where A and B represent species in the binary mixture; x_A and x_B represent molar fractions of adsorbate A and B in the adsorbed phase, respectively; y_A and y_B represent molar fractions of adsorbate A and B in the reservoir, respectively. If the calculated $S_{A/B}$ is less than 1, it indicates that the adsorption capacity of A is lower than that of B (Liu and Wilcox, 2012a,b).

Table 5.1 The basic inputs in the MD simulations.

Mixtures	Molar fraction	Simulation time (ns)	Simulation temperature (K)
Hydrocarbon/CO_2	0.5:0.5	35	363.15
Hydrocarbon/SO_2	0.5:0.5	35	363.15

It is noted that "Hydrocarbon" represents pure C_1, C_2, nC_3, nC_4, or nC_5.

5.2.3 Comparison of the influence of CO_2 with SO_2 on fluid distribution of hydrocarbons in nanopore

Fig. 5.20A shows the density profile of pure C_1 in a 3-nm pore at 363.15 K. Pure C_1 can only form single-adsorption layer on the pore surface, which is in line with the previous findings (Li et al., 2014; Tian et al., 2017). When an equal molar amount of CO_2 is introduced into this 3-nm pore, which is

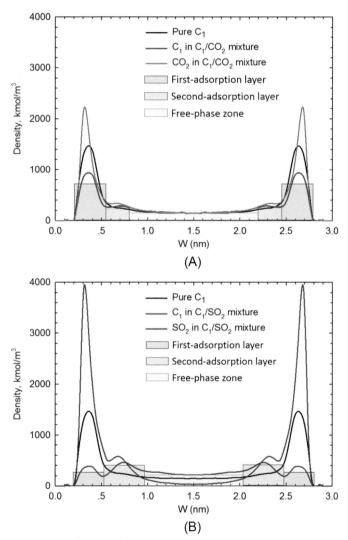

Fig. 5.20 Density distributions of pure C_1, (A) CO_2/C_1, and (B) SO_2/C_1 mixtures in the 3-nm pore at 363.15 K.

initially saturated with C_1, the density of adsorption layer of CO_2 at equilibrium is much higher than that of C_1, indicating the more affinity of CO_2 on organic pore surface. Moreover, density of the adsorption layer of C_1 decreases significantly after CO_2 introduction. CO_2 has a higher adsorption capacity on organic surface than that of C_1; thereby, the adsorbed C_1 can be replaced by CO_2 from pore surface, which thereof decreases the density in the adsorption layer of C_1. As shown in Fig. 5.20A, in addition to the first adsorption layer, C_1 also exhibits a second weak adsorption layer when CO_2 is introduced. Recently, Tian et al. (2017) also observed this second weak adsorption layer but named it by "transition zone," of which the density is significantly smaller than that of the adsorption layer. However, the possible mechanisms forming such zone were not elucidated in their study. With more adsorbed C_1 being replaced by CO_2, the desorbed C_1 can still be attracted by the pore surface as well as the adsorbed molecules in the first adsorption layer, which enhances the formation of the second weak adsorption layer. In other words, the second weak adsorption layer formed by C_1 is resulted from the competitive adsorption between CO_2 and C_1 on pore surface and the attractions from pore surface and first adsorption layer.

Comparatively, an equal molar amount of SO_2 is introduced into the 3-nm pore instead of CO_2, which is initially saturated with C_1. Fig. 5.20B presents the density profiles of pure C_1 and C_1, SO_2 in the equal molar mixture of C_1/SO_2 in the 3-nm pore at 363.15 K. The density of the adsorption layer of SO_2 is significantly higher than that of C_1, indicating the higher adsorption capacity of SO_2 on the organic pore surface. Similarly, the adsorption-layer density of C_1 is reduced heavily after introducing SO_2, suggesting the feasibility of SO_2 in recovering C_1 from organic pores. Compared to CO_2, the adsorption-layer density of SO_2 is much higher. In addition, the density drops in the adsorption layer of C_1 caused by SO_2 is more significant than that caused from CO_2. It suggests that SO_2 can replace the adsorbed C_1 more efficiently than CO_2 from organic surface.

Interestingly, as shown in Fig. 5.20B, density of the second adsorption layer of C_1 is observed to be higher than that of the first adsorption layer after introducing SO_2. It is not proper to recognize the second adsorption layer as the so-called "transition zone" (Tian et al., 2017) only according to the relatively density values. Due to the stronger adsorption capacity of SO_2, SO_2 replaces the adsorbed C_1 from pore surface, releasing more adsorption sites; such released adsorption sites are then immediately occupied by SO_2 molecules. The desorbed C_1 molecules are still highly attracted by the pore surface as well as the first adsorption layer, resulting in the formation of the second stronger adsorption layer adjacent to the first adsorption layer. As shown in Fig. 5.20, after introducing SO$_2$, the density of C$_1$ in the free-phase zone is higher than those after introducing CO$_2$ as well as that of the pure C$_1$; it indicates that more adsorbed C$_1$ can be replaced by SO$_2$ and becomes the free-state C$_1$ in pore center. Thereby, SO$_2$ is possibly more efficient than CO$_2$ for enhancing shale C$_1$ recovery from organic pores.

Besides of C$_1$, heavier components, such as C$_2$, nC$_3$, nC$_4$, and nC$_5$, may also be the important components in shale gas or shale condensate. Figs. 5.21–5.24 show the density profiles of pure C$_2$, nC$_3$, nC$_4$, and nC$_5$ and the equal molar mixtures of CO$_2$/C$_2$, SO$_2$/C$_2$, CO$_2$/nC$_3$, SO$_2$/nC$_3$, CO$_2$/nC$_4$, SO$_2$/nC$_4$, CO$_2$/nC$_5$, and SO$_2$/nC$_5$ in the 3-nm pore at 363.15 K. Similarly, the density of hydrocarbons in the adsorption layer in their binary mixtures is lower than that of the pure hydrocarbons, i.e., C$_2$, nC$_3$, nC$_4$, and nC$_5$. It indicates that CO$_2$ and SO$_2$ can somewhat replace the heavier hydrocarbons from the organic pore surface. However, as hydrocarbon becomes heavier, density of hydrocarbons in the adsorption layer is approaching that of the pure hydrocarbons, suggesting that the heavier hydrocarbons may not be readily to be replaced. It is because the adsorption capacity of CO$_2$ or SO$_2$ relative to hydrocarbons is decreasing when hydrocarbon-species in shale fluids are becoming heavier.

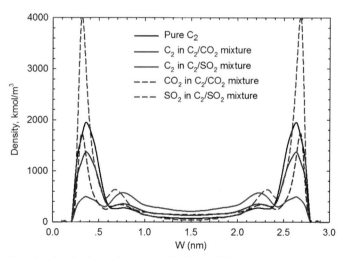

Fig. 5.21 Density distributions of pure C$_2$, CO$_2$/C$_2$, and SO$_2$/C$_2$ mixtures in the 3-nm pore at 363.15 K.

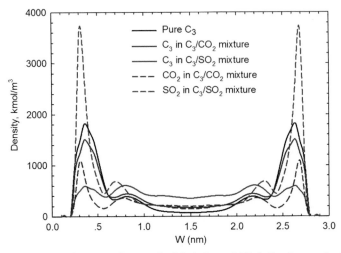

Fig. 5.22 Density distributions of pure nC_3, CO_2/nC_3, and SO_2/nC_3 mixtures in the 3-nm pore at 363.15 K.

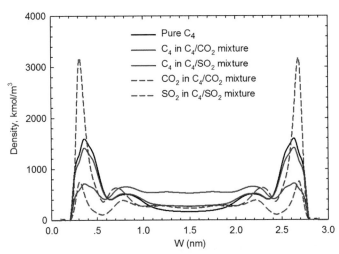

Fig. 5.23 Density distributions of pure nC_4, CO_2/nC_4, and SO_2/nC_4 mixtures in the 3-nm pore at 363.15 K.

In addition, the density of hydrocarbons in adsorption layers drops more significantly after introducing SO_2 than those after CO_2 introduction, indicating the superiority of SO_2 for enhancing the heavier hydrocarbons from organic pores. This observation is in line with the previous founding using the low-field nuclear magnetic resonance technique that SO_2 is more efficient than CO_2 for recovering C_1 and C_2 from shale samples (Huang et al., 2019a,b). Interestingly, the introduction of SO_2 and CO_2 increases the

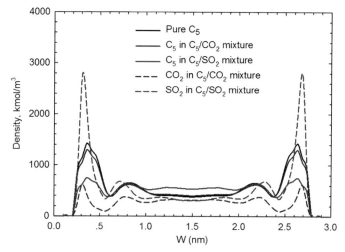

Fig. 5.24 Density distributions of pure nC_5, CO_2/nC_5, and SO_2/nC_5 mixtures in the 3-nm pore at 363.15 K.

density of hydrocarbons in pore center (i.e., the free-phase zone); compared to CO_2, the density in pore center is observed to increase more significantly for SO_2. The desorbed hydrocarbons due to the introduction of SO_2 and CO_2 become free-state gas residing in the pore center; further developing method should be proposed to recover the free-state gas confined in nanopores to enhance the recovery of shale hydrocarbons.

5.2.4 Adsorption selectivity of CO₂ and SO₂ over different hydrocarbon-species

In nanopores, different components in mixtures generally exhibit individual adsorption capacities on organic pore surface, which is defined as the so-called "competitive adsorption." In this work, the adsorption selectivity is calculated to illustrate the relative adsorption capacity of CO_2 or SO_2 over different hydrocarbon-species in their binary mixtures. Fig. 5.25 presents the adsorption selectivity of SO_2 and CO_2 over C_1, C_2, nC_3, nC_4, and nC_5 in their binary mixtures in the 3-nm pore. As shown in Fig. 5.25, the adsorption selectivity of SO_2 over C_1, C_2, nC_3, nC_4, and nC_5 is always higher than that of CO_2 at the same condition; meanwhile, the difference enlarges as hydrocarbon becomes lighter. It proves our aforementioned statement that SO_2 is more affinity to the organic pore surface than CO_2. In addition, as carbon number increases, the adsorption selectivity of SO_2 and CO_2 over hydrocarbons decreases, suggesting that the adsorption capacity of

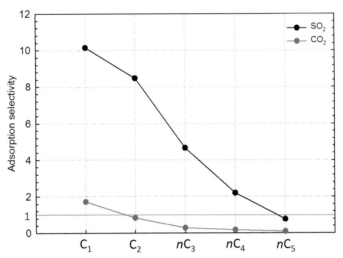

Fig. 5.25 Adsorption selectivity of CO_2 and SO_2 over C_1, C_2, nC_3, nC_4, and nC_5 in their binary mixtures in the 3-nm pore at 363.15 K.

SO_2 and CO_2 relative to hydrocarbons decreases with the increasing carbon number in hydrocarbons.

In Fig. 5.25, adsorption selectivity of SO_2 over C_1, C_2, nC_3, and nC_4 is always higher than 1; it means that the adsorption capacity of SO_2 on organic pore surface is stronger than that of C_1, C_2, nC_3, and nC_4. As for CO_2, the adsorption selectivity is higher than 1 only for C_1, suggesting that the adsorption capacity of CO_2 is higher than that of C_1 but lower than that of C_2, nC_3, nC_4, and nC_5. One may infer that SO_2 is efficient for enhanced C_1, C_2, nC_3, and nC_4 recovery but is not proper for nC_5, while CO_2 is only suitable for the recovery of C_1 from organic pores. However, as has been observed in Fig. 5.24, after introducing SO_2 into the "nC_5-saturated" pore, the adsorption-layer density of nC_5 also decreases slightly, while the density of nC_5 in the free-phase zone increases. Moreover, as shown in Figs. 5.21–5.24, after introducing CO_2, the adsorption-layer density of C_2, nC_3, nC_4, and nC_5 decreases slightly, while the density in the free-phase zone increases. That is, the presence of CO_2 and SO_2 somewhat increases the recovery of C_1, C_2, nC_3, nC_4, and nC_5 from the organic pore due to the effect of competitive adsorption.

5.2.5 Replacement efficiency of CO_2 and SO_2 on different hydrocarbon-species

Hydrocarbons residing in organic pores can be possibly replaced out by the introduced SO_2 or CO_2, resulting in the enhancement of the recovery of hydrocarbons. Fig. 5.26 shows the replacement efficiency of CO_2 or

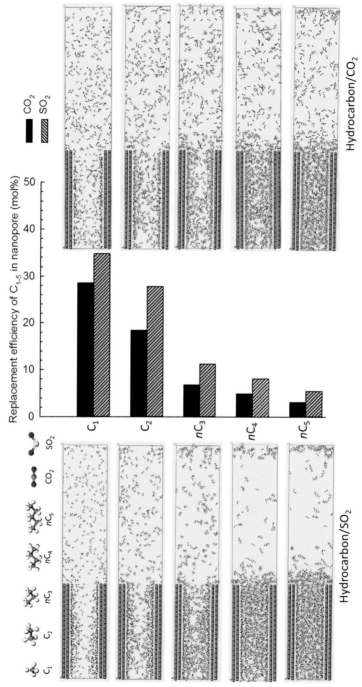

Fig. 5.26 Replacement efficiency of CO_2 and SO_2 on C_1, C_2, nC_3, nC_4, and nC_5 from the 3-nm pore.

SO_2 on different hydrocarbon-species. It is noted that the replacement efficiency is defined as the molar percentage of hydrocarbons replaced by CO_2 or SO_2 from nanopores. The recovery of C_1, C_2, nC_3, nC_4, and nC_5 is enhanced to some extent after introducing CO_2 or SO_2 into the organic pore. Specifically, CO_2 can recover C_1 by more than 25 mol% from the 3-nm pore, while the recovery degree is generally less than 20 mol% for the other hydrocarbon-species. On the contrary, SO_2 can achieve the recovery degree by more than 25 mol% for both C_1 and C_2. Moreover, recovery degree for different hydrocarbon-species achieved by SO_2 is always higher than that of CO_2. It confirms our viewpoint that SO_2 can be a more efficient agent than CO_2 for enhancing hydrocarbon recovery from organic pores. In addition, as hydrocarbon becomes heavier, the recovery achieved by CO_2 and SO_2 decreases; it is mainly caused from the decreasing adsorption capacity of CO_2 or SO_2 relative to hydrocarbon-species.

Fig. 5.26 depicts the snapshots of molecular distributions of C_1, C_2, nC_3, nC_4, nC_5, CO_2 and SO_2 in their binary mixtures in the 3-nm pore at 363.15 K. SO_2 molecules mainly adsorb on pore surface, forming adsorption layers. The adsorbed hydrocarbons, i.e., C_1, C_2, nC_3, nC_4, and nC_5, are replaced by SO_2 molecules, residing in the pore center or in the reservoir. On the contrary, among the five hydrocarbon-species, obvious adsorption layers formed by CO_2 are observed only when CO_2 is introduced into the "C_1-saturated" pore. It seems that CO_2 cannot replace hydrocarbons heavier that C_1; the introduced CO_2 mainly appears in pore center and outside of the organic pore. However, the fact that CO_2 enhances the recovery of C_2, nC_3, nC_4, and nC_5 is probably resulted from the accidently happened adsorption of CO_2 on pore surface when it is introduced to coexist with the heavier hydrocarbons, wherein the adsorption sites of hydrocarbons is partly occupied by the adsorbed CO_2.

This work may provide some fundamental understanding of the mechanisms of enhancing shale hydrocarbon recovery using flue-gas injection method. It is also significant for understanding the basic mechanisms of CO_2 and SO_2 sequestration in shale reservoirs. In future works, SO_2 may also be applied for enhanced coalbed methane recovery besides of CO_2. Moreover, the adsorption behavior is suggested to be studied at varied pressure and temperature conditions, which is practical for designing the developing schemes for shale reservoirs. In shale matrix, pore size varies, showing pore-size distribution; adsorption behavior modeling for porous media with pore-size distribution is different from that in single nanopore. However, the recent works (Liu et al., 2018a,b) mainly investigate adsorption behavior in single pores. Thereby, molecular dynamic simulations are recommended to study adsorption behavior in porous media with pore-size distribution, which is more practical for shale reservoir modeling.

Additionally, based on the scanned SEM imaging, it is found that pore structures in real shale sample are not only slit-shaped, but also include ink-bottle and cylindrical shaped pores (de Boer, and Lippens, 1964; Sing et al., 2008). The adsorption behavior of the hydrocarbon/SO_2 (or CO_2) mixtures is expected to be different from those in the slit-shaped pores. For instance, depending on the pore-diameter ratio of "ink" and "bottle," N_2 adsorption/desorption isotherms may behave differently as the results of pore-blocking and cavitation effects within the ink-bottle model (Fan et al., 2011; Klomkliang et al., 2013). Future work is suggested to explore the effect of pore geometry on the adsorption of nanoconfined hydrocarbon/SO_2 (or CO_2) mixtures. Flue gas is a mixture composed by multiple components. SO_2 may behave differently in flue gas or as a single component. In the future works, it is necessary to investigate the behavior of flue gas for enhancing shale reserve recovery. In addition, besides of kerogen, some other minerals, such as illite, kaolinite, and montmorillonite, may exist. Hydrocarbons generally behave differently on different core minerals. Thereby, it is necessary to investigate the efficiency of CO_2 or SO_2 for recovering hydrocarbons due to the difference in mineral composition.

References

Ambrose, R.J., Hartman, R.C., Diaz-Campos, M., et al., 2012. Shale gas-in-place calculations part I: new pore-scale considerations. SPE J. 17, 219–229.
Ampomah, W., Balch, R.S., Cather, M., Will, R., Gunda, D., Dai, Z., Soltanian, M.R., 2017. Optimal design of CO_2 storage and oil recovery under geological uncertainty. Appl. Energy 195, 80–92.
Andersen, H.C., 1980. Molecular dynamics at constant pressure and/or temperature. J. Chem. Phys. 72, 384–393.
Bachu, S., 2002. Sequestration of CO_2 in geological media in response to climate change: road map for site selection using the transform of the geological space into the CO_2 phase space. Energy Convers. Manag. 43 (1), 87–102.
Bae, J.S., Bhatia, S.K., 2006. High-pressure adsorption of methane and carbon dioxide on coal. Energy Fuel 20 (6), 2599–2607.
Bae, J.S., Bhatia, S.K., Rudolph, V., Massarotto, P., 2009. Pore accessibility of methane and carbon dioxide in coals. Energy Fuels 23 (6), 3319–3327.
Bae, J.S., Nguyen, T.X., Bhatia, S.K., 2014. Pore accessibility of Ti_3SiC_2-derived carbons. Carbon 68, 531–541.
Bagreev, A., Bashkova, S., Bandosz, T.J., 2002. Adsorption of SO_2 onactivated carbons: the effect of nitrogen functionality and pore sizes. Langmuir 18, 1257–1264.
Bhowmik, S., Dutta, P., 2011. Investigation into the methane displacement behavior by cyclic, pure carbon dioxide injection in dry, powdered, bituminous Indian coals. Energy Fuel 25 (2011), 2730–2740.

Brochard, L., Vandamme, M., Pellenq, R.J.M., Fen-Chong, T., 2012. Adsorption-induced deformation of microporous materials: coal swelling induced by CO_2-CH_4 competitive adsorption. Langmuir 28, 2659–2670.

Burnham, A., Han, J., Clark, C.E., Wang, M., Dunn, J.B., Palou-Rivera, I., 2011. Life-cycle greenhouse gas emissions of shale gas, natural gas, coal, and petroleum. Environ. Sci. Technol. 46, 619–627.

Busch, A., Gensterblum, Y., Krooss, B.M., 2003. Methane and CO_2 sorption and desorption measurements on dry Argonne premium coals: pure components and mixtures. Int. J. Coal Geol. 55, 205–224.

Chatterjee, R., Paul, S., 2013. Classification of coal seams for coal bed methane exploitation in central part of Jharia coalfield, India—a statistical approach. Fuel 111, 20–29.

Chávez-Páez, M., Workum, K.V., Pablo, L.D., Pablo, J.J.D., 2001. Monte Carlo simulations of Wyoming sodium montmorillonite hydrates. J. Chem. Phys. 114, 1405–1413.

Chen, M., Kang, Y., Zhang, T., You, L., Li, X., Chen, Z., Wu, K., Yang, B., 2018. Methane diffusion in shales with multiple pore sizes at supercritical conditions. Chem. Eng. J. 334, 1455–1465.

Cheng, A.L., Huang, W.L., 2004. Selective adsorption of hydrocarbon gases on clays and organic matter. Org. Geochem. 35 (4), 413–423.

Crozier, P.S., Rowley, R.L., Spohr, E., Henderson, D., 2000. Comparison of charged sheets and corrected 3D Ewald calculations of long-range forces in slab geometry electrolyte systems with solvent molecules. J. Chem. Phys. 112, 9253–9257.

Curtis, J.B., 2002. Fractured shale-gas systems. AAPG Bull. 86 (11), 1921–1938.

Cygan, R.T., Liang, J.J., Kalinichev, A.G., 2004. Molecular models of hydroxide, oxyhydroxide, and clay phases and the development of a general force field. J. Phys. Chem. B 108, 1255–1266.

Dai, Z., Stauffer, P., Carey, J., Middleton, R., Lu, Z., Jacobs, J., Spangle, L., Hnottavange-Telleen, K., 2014. Pre-site characterization risk analysis for commercial-scale carbon sequestration. Environ. Sci. Technol. 48 (7), 3908–3915.

Dai, Z., Viswanathan, H., Middleton, R., Pan, F., Ampomah, W., Yang, C., Jia, W., Xiao, T., Lee, S., McPherson, B., Balch, R., Grigg, R., White, M., 2016. CO_2 accounting and risk analysis for CO_2 sequestration at enhanced oil recovery sites. Environ. Sci. Technol. 50, 7546–7554.

de Boer, J.H., Lippens, B.C., 1964. Studies on pore systems in catalysts II. The shapes of pores in aluminum oxide systems. J. Catal. 3, 38–43.

Deer, W.A., Howie, R.A., Zussman, J., 1996. An Introduction to the Rock Forming Minerals. Longman, London.

Dong, Y., Li, Y., Zhang, L., Cui, L., Zhang, B., Dong, Y., 2017. Novel method of ultralow SO_2 emission for CFB boilers: combination of limestone injection and activated carbon adsorption. Energy Fuel 31, 11481–11488.

Duan, S., Gu, M., Du, X., Xian, X., 2016. Adsorption equilibrium of CO_2 and CH_4 and their mixture on Sichuan Basin shale. Energy Fuel 30 (3), 2248–2256.

Etminan, S.R., Javadpour, F., Maini, B.B., Chen, Z., 2014. Measurement of gas storage processes in shale and of the molecular diffusion coefficient in kerogen. Int. J. Coal Geol. 123, 10–19.

Faiz, M.M., et al., 2007. Evaluating geological sequestration of CO_2 in bituminous coals: the southern Sydney Basin, Australia as a natural analogue. Int. J. Greenhouse Gas Control 1, 223–235.

Falk, K., Coasne, B., Pellenq, R., Ulm, F.J., Bocquet, L., 2015. Subcontinuum mass transport of condensed hydrocarbons in nanoporous media. Nat. Commun. 6, 6949.

Fan, C., Do, D.D., Nicholson, D., 2011. On the cavitation and pore blocking in slit-shaped ink-bottle pores. Langmuir 27 (2011), 3511–3526.

Gensterblum, Y., Busch, A., Kroos, B.M., 2014. Molecular concept and experimental evidence of competitive adsorption of H_2O, CO_2 and CH_4 on organic material. Fuel 115, 581–588.

Godec, M., et al., 2013. Potential for enhanced gas recovery and CO_2 storage in the Marcellus shale in the Eastern United States. Int. J. Coal Geol. 118, 95–104.

Heller, R., Zoback, M., 2014. Adsorption of methane and carbon dioxide on gas shale and pure mineral samples. J. Unconven. Oil Gas Resour. 8, 14–24.

Huang, L., Ning, Z., Wang, Q., Qi, R., Zeng, Y., Qin, H., Ye, H., Zhang, W., 2018a. Molecular simulation of adsorption behaviors of methane, carbon dioxide and their mixtures on kerogen: effect of kerogen maturity and moisture content. Fuel 211, 159–172.

Huang, L., Ning, Z., Wang, Q., Zhang, W., Cheng, Z., Wu, X., Qin, H., 2018b. Effect of organic type and moisture on CO_2/CH_4 competitive adsorption in kerogen with implication. Appl Energy 210, 28–43.

Huang, X., Li, T., Gao, H., Zhao, J., Wang, C., 2019a. Comparison of SO_2 with CO_2 for recovering shale resources using low-field nuclear magnetic resonance. Fuel 245, 563–569.

Huang, X., Li, A., Li, X., Liu, Y., 2019b. Influence of typical core minerals on tight oil recovery during CO_2 flooding using the nuclear magnetic resonance technique. Energy Fuel 33, 7147–7154.

Hughes, J.D., 2013. Energy: a reality check on the shale revolution. Nature 494, 307–308.

Ji, L., Zhang, T., Milliken, K.L., Qu, J., Zhang, X., 2012. Experimental investigation of main controls to methane adsorption in clay-rich rocks. Appl. Geochem. 27 (12), 2533–2545.

Jiang, X., 2011. A review of physical modelling and numerical simulation of long-term geological storage of CO_2. Appl. Energy 88 (11), 3557–3566.

Jiang, J., Shao, Y., Younis, R.M., 2014. Development of a multi-continuum multicomponent model for enhanced gas recovery and CO_2 storage in fractured shale gas reservoirs. In: Proceedings of the SPE improved oil recovery symposium. Society of Petroleum Engineers, Tulsa, Oklahoma, April 12–16, 2014. SPE 169114.

Jin, Z., Firoozabadi, A., 2014. Effect of water on methane and carbon dioxide sorption in clay minerals by Monte Carlo simulations. Fluid Phase Equilib. 382, 10–20.

Jones, J.E., 1924. On the determination of molecular field-OII from the equation of state of a gas. Proc. R. Soc. Lond. Ser. A 106, 463–477.

Juray, D.W., Guy, B.M., 2000. Investigation of simultaneous adsorption of SO_2 and NO_X on Na-γ-alumina with transient techniques. Catal. Today 62, 319–328.

Kadoura, A., Nair, A.K.N., Sun, S., 2016. Adsorption of carbon dioxide, methane, and their mixture by montmorillonite in the presence of water. Microporous Mesoporous Mater. 225, 331–341.

Kang, S.M., Fathi, E., Ambrose, R.J., Akkutlu, I.Y., Sigal, R.F., 2010. Carbon dioxide storage capacity of organic-rich shales. SPE J. 16 (4), 842–855.

Karacan, C.Ö., et al., 2011. Coal mine methane: a review of capture and utilization practices with benefits to mining safety and to greenhouse gas reduction. Int. J. Coal Geol. 86, 121–156.

Kazemi, M., Takbiri-Borujeni, A., 2016. Molecular dynamics study of carbon dioxide storage in carbon-based organic nanopores. In: Presented in the SPE Annual Technical Conference and Exhibition, 26–28 September, Dubai, UAE. SPE 181705.

Khosrokhavar, R., Wolf, K.H., Bruining, H., 2014. Sorption of CH_4 and CO_2 on a carboniferous shale from Belgium using a manometric setup. Int. J. Coal Geol. 128, 153–161.

Kim, T.H., Cho, J., Lee, K.S., 2017. Evaluation of CO_2 injection in shale gas reservoirs with multi-component transport and geomechanical effects. Appl. Energy 190, 1195–1206.

Klomkliang, N., Do, D.D., Nicholson, D., 2013. On the hysteresis loop and equilibrium transition in slit-shaped ink-bottle pores. Adsorption 19 (2013), 1273–1290.

Kowalczyk, P., et al., 2012. Displacement of methane by coadsorbed carbon dioxide is facilitated in narrow carbon nanopores. J. Phys. Chem. C 116, 13640–13649.

Kurniawan, Y., Bhatia, S.K., Rudolph, V., 2006. Simulation of binary mixture adsorption of methane and CO_2 at supercritical conditions in carbons. AICHE J. 52 (3), 957–967.

Lee, J.S., Kim, J.H., Kim, J.T., 2002. Adsorption equilibria of CO_2 on zeolite 13X and zeolite X/activated carbon composite. J. Chem. Eng. Data 47, 1237–1242.

Li, Z., Jin, Z., Firoozabadi, A., 2014. Phase behavior of adsorption of pure substances and mixtures and characterization in nanopore structures by density functional theory. SPE J. 19, 1096–1109.

Li, Y., Wang, S., Wang, Q., et al., 2016. Molecular dynamics simulations of tribology properties of NBR (nitrile-butadiene rubber)/carbon nanotube composites. Compos. Part B 97, 62–67.

Liu, Y., Hou, J., 2019. Investigation on the potential relationships between geophysical properties and CH_4 adsorption in a typical shale gas reservoir. Energy Fuel 33, 8354–8362.

Liu, Y., Wilcox, J., 2012a. Molecular simulation studies of CO_2 adsorption by carbon model compounds for carbon capture and sequestration applications. Environ. Sci. Technol. 47, 95–101.

Liu, Y., Wilcox, J., 2012b. Molecular simulations of CO_2 adsorption in micro- and mesoporous carbons with surface heterogeneity. Int. J. Coal Geol. 104, 83–95.

Liu, Q.Y., Liu, Z.Y., Wu, W.Z., 2009. Effect of V_2O_5 additive on simultaneous SO_2 and NO removal from flue gas over a monolithic cordierite-based CuO/Al_2O_3 catalyst. Catal. Today 147S, S285–S289.

Liu, F., et al., 2013. Assessing the feasibility of CO_2 storage in the New Albany shale (Devonian-Mississippian) with potential enhanced gas recovery using reservoir simulation. Int. J. Greenhouse Gas Control 17, 111–126.

Liu, X., He, X., Qiu, N., Yang, X., Tian, Z.Y., Li, M., Xue, Y., 2016. Molecular simulation of CH_4, CO_2, H_2O and N_2 molecules adsorption on heterogeneous surface models of coal. Appl. Surf. Sci. 389, 894–905.

Liu, Y., Jin, Z., Li, H., 2018a. Comparison of Peng-Robinson equation of state with capillary pressure model with engineering density-functional theory in describing the phase behavior of confined hydrocarbons. SPE J. 23 (05), 1784–1797.

Liu, Y., Li, H.A., Tian, Y., Jin, Z., Deng, H., 2018b. Determination of the absolute adsorption/desorption isotherms of CH_4 and n-C_4H_{10} on shale from a nano-scale perspective. Fuel 218, 67–77.

Lopez, D., Buitrago, R., Escribano, S.A., Reinoso, R.F., Mondragon, F., 2007. Surface complexes formed during simultaneous catalytic adsorption of NO and SO_2 on activated carbons at low temperatures. J. Phys. Chem. C 111, 1417–1423.

Lorentz, H.A., 1881. Nachtrag zu der Abhandlung: Ueber die Anwendung des Satzes vom Virial in der kinetischen Theorie der Gase. Ann. Phys. 248 (4), 660–661.

Lu, X.C., Li, F.C., Watson, A.T., 1995. Adsorption measurements in Devonian shales. Fuel 74 (4), 599–603.

Lu, X., Jin, D., Wei, S., Zhang, M., Zhu, Q., Shi, X., Deng, Z., Guo, W., Shen, W., 2015. Competitive adsorption of a binary CO_2-CH_4 mixture in nanoporous carbons: effects of edge-functionalization. Nanoscale 7 (3), 1002–1012.

Luo, F., Xu, R.N., Jiang, P.X., 2013. Numerical investigation of the influence of vertical permeability heterogeneity in stratified formation and of injection/production well perforation placement on CO_2 geological storage with enhanced CH_4 recovery. Appl. Energy 102, 1314–1323.

Luo, L., Guo, Y., Zhu, T., Zheng, Y., 2017. Adsorption species distribution and multicomponent adsorption mechanism of SO_2, NO, and CO_2 on commercial adsorbents. Energy Fuel 31, 11026–11033.

Mac Dowell, N., Fennell, P.S., Shah, N., Maitland, G.C., 2017. The role of CO_2 capture and utilization in mitigating. Nat. Clim. Chang. 7, 243–249.

Majewska, Z., et al., 2009. Binary gas sorption/desorption experiments on a bituminous coal: simultaneous measurements on sorption kinetics, volumetric strain and acoustic emission. Int. J. Coal Geol. 77, 90–102.

Martin, M.G., Siepmann, J.I., 1998. Transferable potentials for phase equilibria.1. United-atom description of n-alkanes. J. Phys. Chem. B 102 (14), 2569–2577.

Montgomery, S.L., Jarvie, D.M., Bowker, K.A., Pollastro, R.M., 2005. Mississippian Barnett Shale, Fort Worth basin, north-central Texas: gas-shale play with multi-trillion cubic foot potential. AAPG Bull. 89, 155–175.

Nguyen, T.X., Bhatia, S.K., 2007. Determination of pore accessibility in disordered nanoporous materials. J. Phys. Chem. C 111 (5), 2212–2222.

Nguyen, T.X., Bhatia, S.K., 2008. Kinetic restriction of simple gases in porous carbons: transition-state theory study. Langmuir 24 (1), 146–154.

Nuttall, B.C., Eble, C.F., Drahovzal, J.A., Bustin, R.M., 2005. Analysis of Devonian black shales in Kentucky for potential carbon dioxide sequestration and enhanced natural gas production. In: Report Kentucky Geological Survey/University of Kentucky.

Ortiz Cancino, O., Pérez, D.P., Pozo, M., Bessieres, D., 2017. Adsorption of pure CO_2 and a CO_2/CH_4 mixture on a black shale sample: manometry and microcalorimetry measurements. J. Pet. Sci. Eng. 159, 307–313.

Ottiger, S., et al., 2008. Measuring and modeling the competitive adsorption of CO_2, CH_4, and N_2 on a dry coal. Langmuir 24, 9531–9540.

Psarras, P., Holmes, R., Vishal, V., Wilcox, J., 2017. Methane and CO_2 adsorption capacities of kerogen in the Eagle Ford shale from molecular simulation. Acc. Chem. Res. 50, 1818–1828.

Rigby, D., Sun, H., Eichinger, B.E., 1997. Computer simulations of poly (ethylene oxide): force field, PVT diagram and cyclization behaviour. Polym. Int. 44, 311–330.

Ross, D.J., Bustin, R.M., 2007a. Impact of mass balance calculations on adsorption capacities in microporous shale gas reservoirs. Fuel 86, 2696–2706.

Ross, D.J.K., Bustin, R.M., 2007b. Shale gas potential of the lower Jurassic Gordondale member, northeastern British Columbia, Canada. Bull. Can Pet. Geol. 55 (1), 51–75.

Ross, D.J.K., Bustin, R.M., 2009. The importance of shale composition and pore structure upon gas storage potential of shale gas reservoirs. Mar. Petrol. Geol. 26 (6), 916–927.

Sainz-Diaz, C.I., Palin, E.J., Dove, M.T., Hernandez-Laguna, A., 2003. Monte Carlo simulations of ordering of Al, Fe, and Mg cations in the octahedral sheet of smectites and illites. Am. Mineral. 88 (7), 1033–1045.

Sing, K.S.W., Everett, D.H., Haul, R.A.W., et al., 2008. Reporting physisorption data for gas/solid systems. In: Handbook of Heterogeneous Catalysis. Wiley-VCH Verlag GmbH & Co. KGaA.

Siriwardane, R.V., Shen, M.S., Fisher, E.P., Poston, J.A., 2001. Adsorption of CO_2 on molecular sieves and activated carbon. Energy Fuel 15, 279–284.

Song, Y., Jiang, B., Li, W., 2017. Molecular simulations of $CH_4/CO_2/H_2O$ competitive adsorption on low rank coal vitrinite. Phys. Chem. Chem. Phys. 19, 17773–17788.

Sumathi, S., Bhatia, S., Lee, K.T., Mohamed, A.R., 2010a. Selection of best impregnated palm shell activated carbon (PSAC) for simultaneous removal of SO_2 and NO_X. J. Hazard. Mater. 176, 1093–1096.

Sumathi, S., Bhatia, S., Lee, K.T., Mohamed, A.R., 2010b. Adsorption isotherm models and properties of SO_2 and NO removal by palm shell activated carbon supported with cerium (Ce/PSAC). Chem. Eng. J. 162, 194–200.

Sun, H., Zhao, H., Qi, N., Li, Y., 2017. Molecular insights into the enhanced shale gas recovery by carbon dioxide in kerogen slit nanopores. J. Phys. Chem. C 121 (18), 10233–10241.

Szczerba, M., Derkowski, A., Kalinichev, A.G., Środoń, J., 2015. Molecular modeling of the effects of 40Ar recoil in illite particles on their K-Ar isotope dating. Geochim. Cosmochim. Acta 159, 162–176.

Tenney, C.M., Cygan, R.T., 2014. Molecular simulation of carbon dioxide, brine, and clay mineral interactions and determination of contact angles. Environ. Sci. Technol. 48, 2035–2042.

Tian, Y., Yan, C., Jin, Z., 2017. Characterization of methane excess and absolute adsorption in various clay nanopores from molecular simulation. Sci. Rep. 7, 12040.

Valentini, P., Schwartzentruber, T.E., Cozmuta, I., 2011. ReaxFF grand canonical Monte Carlo simulation of adsorption and dissociation of oxygen on platinum (111). Surf. Sci. 605, 1941–1950.

Vieira, R.B., Moura, P.A.S., Vilarrasa-García, E., Azevedo, D.C.S., Pastore, H.O., 2018. Polyamine-grafted Magadiite: high CO_2 selectivity at capture from CO_2/N_2 and CO_2/CH_4 mixtures. J. CO_2 Util. 23, 29–41.

Vishal, V., Singh, T.N., Ranjith, P.G., 2015. Influence of sorption time in CO_2-ECBM process in Indian coals using coupled numerical simulation. Fuel 139, 51–58.

Wang, H., Wang, X., Jin, X., Cao, D., 2016a. Molecular dynamics simulation of diffusion of shale oils in montmorillonite. J. Phys. Chem. C 120, 8986–8991.

Wang, X., Zhai, Z., Jin, X., Wu, S., Li, J., Sun, L., Liu, X., 2016b. Molecular simulation of CO_2/CH_4 competitive adsorption in organic matter pores in shale under certain geological conditions. Pet. Explor. Dev. 43 (5), 841–848.

Wardle, R., Brindley, G.W., 1972. Crystal-structures of pyrophyllite, 1Tc, and of its dehydroxylate. Am. Mineral. 57 (5–6), 732–750.

Weijermars, R., 2013. Economic appraisal of shale gas plays in continental Europe. Appl. Energy 106, 100–115.

Weijermars, R., 2014. US shale gas production outlook based on well roll-out rate scenarios. Appl. Energy 124, 283–297.

White, C.M., Smith, D.H., Jones, K.L., Goodman, A.L., Jikich, S.A., LaCount, R.B., DuBose, S.B., Ozdemir, E., Morsi, B.I., Schroeder, K.T., 2005. Sequestration of carbon dioxide in coal with enhanced coalbed methane recovery a review. Energy Fuel 19 (3), 659–724.

Wu, K., Chen, Z., Li, X., 2015a. Real gas transport through nanopores of varying cross-section type and shape in shale gas reservoirs. Chem. Eng. J. 281, 813–825.

Wu, H., Chen, J., Liu, H., 2015b. Molecular dynamics simulations about adsorption and displacement of methane in carbon nanochannels. J. Phys. Chem. C 119, 13652–13657.

Yamazaki, T., Aso, K., Chinju, J., 2006. Japanese potential of CO_2 sequestration in coal seams. Appl. Energy 83, 911–920.

Yang, N., Liu, S., Yang, X., 2015. Molecular simulation of preferential adsorption of CO_2 over CH_4 in Na-montmorillonite clay material. Appl. Surf. Sci. 356, 1262–1271.

Yi, H., Wang, Z., Liu, H., Tang, X., Ma, D., Zhao, S., Zhang, B., Gao, F., Zuo, Y., 2014. Adsorption of SO_2, NO, and CO_2 on activated carbons: equilibrium and thermodynamics. J. Chem. Eng. Data 59, 1556–1563.

Yu, W., Al-Shalabi, E.W., Sepehrnoori, K., 2014. A sensitivity study of potential CO_2 injection for enhanced gas recovery in Barnett shale reservoirs. In: Proceedings of the SPE unconventional resources conference. Society of Petroleum Engineers, Woodlands, Texas, April 1–3, SPE 169012.

Yu, S., Yanming, Z., Wu, L., 2017. Macromolecule simulation and CH_4 adsorption mechanism of coal vitrinite. Appl. Surf. Sci. 396, 291–302.

Yuan, J., Luo, D., Feng, L., 2015a. A review of the technical and economic evaluation techniques for shale gas development. Appl. Energy 148, 49–65.

Yuan, Q., Zhu, X., Lin, K., Zhao, Y.P., 2015b. Molecular dynamics simulations of the enhanced recovery of confined methane with carbon dioxide. Phys. Chem. Chem. Phys. 17, 31887–31893.

Zhang, T., Ellis, G.S., Ruppel, S.C., Milliken, K., Yang, R., 2012. Effect of organic-matter type and thermal maturity on methane adsorption in shale-gas systems. Org. Geochem. 47, 120–131.

Zhang, J., Liu, K., Clennell, M.B., Dewhurst, D.N., Pervukhina, M., 2015. Molecular simulation of CO_2-CH_4 competitive adsorption and induced coal swelling. Fuel 160, 309–317.

Zhao, Y., Feng, Y., Zhang, X., 2016. Molecular simulation of CO_2/CH_4 self- and transport diffusion coefficients in coal. Fuel 165, 19–27.

Ziołek, M.I., Nowak, I.S., Daturi, M., 2000. Effect of sulfur dioxide on nitric oxide adsorption and decomposition on Cu-containing micro- and mesoporous molecular sieves. Top. Catal. 11, 343–350.

CHAPTER SIX

Summary and commendations

6.1 Summary of this book

The understanding of adsorption behavior, phase, and interfacial properties of fluid in shale reservoirs is significant in enhanced shale resources recovery and hydrocarbon-in-place estimation. This thesis investigates the adsorption, fluid-phase behavior, and interfacial tension of fluid at the pore scale. First, density functional theory (DFT) and the PR-EOS with capillary pressure model are employed to investigate phase behavior of pure hydrocarbon and hydrocarbon mixtures in single nanopores. An explicit comparison is then conducted between the statistical thermodynamic-based method and the PR-EOS with capillary pressure model on the phase behavior of confined fluids. Then, molecular dynamics (MD) simulation is applied to investigate the phase behavior of hydrocarbons in porous media, taking into consideration the effect of pore-size distribution. By realizing that the fluid/pore surface interactions can be strong in organic pores, we measure the excess adsorption isotherms of pure C_1 and pure nC_4 on shale samples; then, the absolute adsorption isotherms are obtained by correcting the excess adsorption isotherms with the calculated adsorption-phase density, while such density values are computed using grand canonical Monte Carlo (GCMC) method. Next, an experimental approach is designed to measure the fluid-phase behavior with the presence of shale samples, demonstrating the effect of competitive adsorption on phase behavior. Finally, considering that interfacial tension (IFT) between two phases may also influence phase properties, we measure the IFT of $CO_2/CH_4/$brine system under reservoir conditions using asymmetric drop shape analysis (ADSA) method.

6.2 Suggested future work

(1) Because the PR-EOS with capillary pressure model neglects the interactions between pore surface and molecules, it cannot reliably describe fluid-phase behavior in confined space. New terms taking into consideration such interactions are expected to combine with the PR–EOS with capillary pressure model.

(2) The adsorption behavior of hydrocarbons is investigated in organic porous media using MD simulations. In real shale samples, clay minerals may also take up a large proportion; thus, adsorption behavior of hydrocarbons should also be conducted on mineral nanopores, such as montmorillonite, illite, and kaolinite.

(3) Due to the limitations of the thermogravimetric analysis (TGA) setup, the operating pressure for $n\text{-}C_4H_{10}$ is only up to 5 bar. In the future work, the testing pressure for $n\text{-}C_4H_{10}$ should be higher to simulate the conditions in shale reservoirs.

Index

Note: Page numbers followed by *f* indicate figures, and *t* indicate tables.

A

ADSA. *See* Axisymmetric drop shape analysis (ADSA) method
Adsorption/desorption isotherms, of CH_4 and $n\text{-}C_4H_{10}$ on shale, 74–101
  absolute isotherms, 93–98, 94–95*f*, 97*f*
  adsorption phase, average density of, 88–92, 89–91*f*, 92*t*, 93*f*
  density distributions in nanopores, 84–89
  excess adsorption/desorption isotherms, 78–81, 80*f*
  grand canonical Monte Carlo (GCMC) simulations, 81–84
    vs. conventional approach, 98–101, 99–100*f*
  pore-size distributions, 78, 79*f*, 87–88, 87*f*
  shale sample, characterization of, 76–78, 76*t*, 77*f*, 79*f*
  system pressure, effect of, 84–85, 85*f*
  system temperature, effect of, 85–87, 86*f*
  TOC and BET surface areas, 76*t*
Adsorption selectivity
  of CO_2/CH_4
    illite, 201–202, 202*f*
    kaolinite, 201–202, 203*f*
    montmorillonite, 201–202, 202*f*
  of species in organic pores, 70–71, 71–72*f*
Autosorb iQ-Chemiadsorption, 38, 78, 124–125
Average absolute relative error (AARE), 174–176
Axisymmetric drop shape analysis (ADSA) method, 156–159, 156–157*f*, 158*t*

B

Binary interaction coefficient, 12–13
Boyle-Charles law, 40
Brunauer-Emmett-Teller (BET) method, 57–58, 78, 104

Bubble-point pressures, 35–36, 45–47, 46*t*
Bulk pressure, 15, 17–19, 23

C

Capillary condensation, 16–19, 28–31
Capillary effect, 2, 9–10, 17–19
Capillary pressure, 2, 9–10, 13–14
Carbon-slit pore, 209–210
CCE. *See* Constant composition expansion (CCE) method
CH_4-solid surface interaction, 107–108
CLAYFF force field, 192
CO_2-based fracturing technique, 151
CO_2/CH_4/brine systems, interfacial tension (IFT) for, 4–6
  ADSA method, 156–159, 156–157*f*, 158*t*
  CO_2 concentration, 167–172, 168–171*f*, 173*f*
  effect of pressure, temperature and salinity, 162–167, 163–164*f*
  existing correlations, comparison with, 176, 177–179*f*
  improved IFT model for, 173–179, 175*f*, 176*t*, 180*t*, 180*f*
  mathematical formulation, 159–162
Competitive adsorption behavior, 4
  of C_1/CO_2 mixture, 64, 65*f*
  of $C_1/nC_4/CO_2$ mixture, 68, 69*f*
  of C_1/nC_4 mixture, 62, 63*f*
  of hydrocarbons and hydrocarbon/CO_2 mixtures, 59–74
    adsorption selectivity of species in organic pores, 70–71, 71–72*f*
    C_1 and nC_4 replacement from nanopores with CO_2 injection, 72–74, 73*f*
    carbon sheet, 60–61, 61*f*
    double-nanopore system, 60–70, 60*f*

Competitive adsorption behavior
    (Continued)
    molecular dynamic simulation, 59–62,
        62t
    simulation model, 60–62
    of nC_4/CO_2 mixture, 66, 67–68f
Condensed-phased-optimized molecular
        potential for atomistic
        simulation studies (COMPASS)
        force field, 59–60, 208–209
Confined fluid-phase behavior, in shale
    C_1–nC_6 mixture, 22–35, 24–27f, 29–32f,
        34f
    density functional theory (DFT), 11–13,
        12t
    dew-point calculation, 15
    molecular model and theory, 11–15
    N_2/n-C_4H_{10}, 35–52
    PR-EOS with capillary pressure model,
        13–15
    pure hydrocarbons, critical properties of,
        15–16
    pure nC_8, phase behavior and critical
        properties of, 16–22, 17–20f, 22f
Confined spaces, 35–36
Constant composition expansion (CCE)
        method, 38, 41–42, 47
CO_2 sequestration, 201–205, 202–205f
Cricondentherm point, of confined fluids,
        32
Critical temperature, of lower dew-point,
        23–28, 27f

D

Degassing and drying treatment, 38
Density functional theory (DFT), 10–13,
        12t, 227
Density gradient theory (DGT), 5–6
Desorption isotherms, 28–31, 32f
Dew-point calculation, 15
Dew-point temperature, of confined fluids,
        16–17
DFT. *See* Density functional theory (DFT)
Differential scanning calorimetry,
        36–37
Double-nanopore system, 60–70, 60f

E

Energy-dispersive X-ray spectroscopy
        (EDX) analysis, 76–78, 77f, 104–106
Euler-Lagrange equation, 12

G

Gas chromatography (GC), 41–42
Gas mixtures, phase behavior of, 4
Grand canonical Monte Carlo (GCMC)
        simulations, 2–3, 10, 227
    absolute adsorption isotherms of CH_4 on
        shale, low-field NMR technique,
        130–132
    adsorption/desorption isotherms of
        CH_4 and n-C_4H_{10} on shale from
        nanoscale, 81–84
    selective adsorption of $CO_2/$
        CH_4 mixture on clay-rich shale,
        192–195, 194–195f
Grand potential functional, 11–12

H

Helmholtz free energy functional, 11–12
Hydrocarbons
    competitive adsorption behavior of,
        59–74
        adsorption selectivity of species in
            organic pores, 70–71, 71–72f
        C_1 and nC_4 replacement from
            nanopores with CO_2 injection,
            72–74, 73f
        carbon sheet, 60–61, 61f
        double-nanopore system, 60–70, 60f
        molecular dynamic simulation, 59–62,
            62t
        simulation model, 60–62
    confinement-induced supercriticality, 28
    pure, adsorption behavior on shale, 3–4

I

Ideal adsorbed solution theory (IAST),
        57–58
IFT. *See* Interfacial tension (IFT)
Illite
    adsorption capacity of CO_2/CH_4 mixture
        on, 194–195, 195f

adsorption selectivity of CO_2/CH_4,
201–202, 202*f*
density profiles of CO_2/CH_4, 196–201,
196*f*, 198*f*, 200*f*
structure, 190, 191*f*
Interfacial tension (IFT), 9–10, 227
for CO_2/CH_4/brine systems, 4–6
ADSA IFT apparatus, 156–159,
156–157*f*, 158*t*
CO_2 concentration, 167–172,
168–171*f*, 173*f*
effect of pressure, temperature and
salinity, 162–167, 163–164*f*
existing correlations, comparison with,
176, 177–179*f*
improved IFT model for, 173–179,
175*f*, 176*t*, 180*t*, 180*f*
mathematical formulation, 159–162
of pure component, 14
reduction effect, 167

K

Kaolinite
adsorption capacity of CO_2/CH_4 mixture
on, 194–195, 195*f*
adsorption selectivity of CO_2/CH_4,
201–202, 203*f*
density profiles of CO_2/CH_4, 196–201,
197*f*, 199–200*f*
structure, 191, 191*f*
Kelvin equation, 2, 9–10

L

Lab-on-a-chip technology, 36–37
Langmuir adsorption model, 57–58
Linear gradient theory, 5–6
Lorentz-Berthelot combining rules, 12–13,
192
Low-field nuclear magnetic resonance
(NMR) technique, absolute
adsorption isotherms of CH_4 on
shale with, 122–141
absolute adsorption, 127
absolute adsorption isotherms, 136–137,
136*f*
adsorption-phase density, 133–136,
134–135*f*

excess adsorption isotherms, 132–133,
133*f*
excess adsorption, measurements of,
125–127
grand canonical Monte Carlo (GCMC)
simulations, 130–132
NMR test, 128–130, 128*t*, 128*f*
shale samples, characterization of,
124–125, 124–126*f*
T_2 spectrum, 137–139, 138*f*

M

Modified Buckingham exponential-6
intermolecular potential, 13, 81–82
Molecular clay-mineral models, 190–191,
191*f*
Molecular dynamic (MD) simulation, 227
comparing effectiveness of SO_2 with
CO_2 for replacing hydrocarbons
from nanopores, 208–209, 210*t*
competitive adsorption behavior of
hydrocarbons and hydrocarbon/
CO_2 mixtures, 59–62, 62*t*
Monte Carlo (MC) simulations, 20–22, 35
Montmorillonite
adsorption capacity of CO_2/CH_4 mixture
on, 194–195, 194–195*f*
adsorption selectivity of CO_2/CH_4,
201–202, 202*f*
density profiles of CO_2/CH_4, 196–197,
196*f*, 199–201, 199–200*f*, 211*f*
structure, 190–191, 191*f*

N

N_2 adsorption/desorption test, 38–40, 43*f*
Nanopores
adsorbed CH_4 density in, 115, 116*f*
comparing effectiveness of SO_2 with
CO_2 for replacing hydrocarbons,
206–219
adsorption selectivity, 215–216, 216*f*
fluid distribution, 211–215, 211*f*,
213–215*f*
molecular dynamics, 208–209, 210*t*
replacement efficiency, 216–219, 217*f*
simulation model, 209–210
confined fluid-phase behavior in, 2–3

Nanopores *(Continued)*
  density distributions of CH_4 in, 84–89,
      111–115, 111–114*f*
National Institute of Standards and
    Technology (NIST), 14, 21, 41–42,
    193
Negative flash algorithm, 14–15
NIST. *See* National Institute of Standards
    and Technology (NIST)
Nitrile O-rings, 156
N_2/n-C_4H_{10}, phase behavior of, 35–52
  bubble-point pressure, 45–47, 46*t*
  compositions and molar numbers, 40–41,
      41*t*
  mass, pore volume, and TOC of shale
      samples, 39, 40*t*
  in partially confined space, 44–47, 44–46*f*
  pore size distribution, 43–44, 43*f*
  P/V isotherms, measurement of, 38–42,
      39*f*, 44–45, 44–46*f*
  sorption of individual components, 47–49
  TOC on sorption capacity, 49–52, 50*t*,
      51*f*
Nonlocal density functional theory
    (NLDFT), 78, 104

P

Parachor model, 14, 159–160
Peng-Robinson equation of state
    (PR-EOS), 2, 9–10, 108, 158–159
Peng-Robinson equation of state (PR-EOS)
    with capillary pressure model, 227
  vs. engineering density functional theory,
      11–35
  fluid distribution in nanopore, 9–10
  fluid-surface interactions, 9–10
Perturbation theory, 5–6
Phase behavior
  of confined C_1-nC_6 mixture, 22–35,
      24–27*f*, 29–32*f*, 34*f*
  of confined pure nC_8, 16–22, 17–20*f*, 22*f*
  of N_2/n-C_4H_{10}, 35–52
    bubble-point pressure, 45–47, 46*t*
    compositions and molar numbers,
        40–41, 41*t*
    mass, pore volume, and TOC of shale
        samples, 39, 40*t*
    in partially confined space, 44–47,
        44–46*f*

  pore size distribution, 43–44, 43*f*
  P/V isotherms, measurement of,
      38–42, 39*f*, 44–45, 44–46*f*
  sorption of individual components,
      47–49
  TOC on sorption capacity, 49–52, 50*t*,
      51*f*
  of pure hexane, heptane, and octane,
      36–37
Physi-adsorption Gas Adsorption Analyzer,
    38, 78, 124–125
Propane, dew-point pressure of, 19–20
Pure hydrocarbons
  adsorption behavior of, 3–4
  critical properties of, 15–16

R

Rectilinear law, 16
Resistance temperature device (RTD)
    sensor, 156

S

Selective adsorption of CO_2/CH_4 mixture,
    on clay-rich shale using molecular
    simulations
  CH_4 recovery and CO_2 sequestration,
      201–205, 202–205*f*
  force field parameters, 192
  GCMC simulations, 192–195, 194–195*f*
  molecular clay-mineral models, 190–191,
      191*f*
  pore size, effect of, 199–201, 200*f*
  system pressure, effect of, 196–198,
      196–197*f*
  system temperature, effect of, 198,
      198–199*f*
Shale
  absolute adsorption of CH_4 on, 101–121
    absolute adsorption isotherms from
        SLD model, 118–121, 118–119*t*,
        120–121*f*
    adsorbed CH_4 density in nanopores,
        115, 116*f*
    density distributions, 111–115,
        111–114*f*
    low-field nuclear magnetic resonance,
        122–141
    shale samples, characterization of,
        104–106, 105–106*f*, 106*t*

simplified local density (SLD) theory,
    106–111
validation of SLD model, 115–118,
    116–117*f*
adsorption behavior of pure
    hydrocarbons, 3–4
adsorption/desorption isotherms of
    CH₄ and *n*-C₄H₁₀ on, 74–101
absolute isotherms, 93–98, 94–95*f*, 97*f*
adsorption phase, average density of,
    88–92, 89–91*f*, 92*t*, 93*f*
density distributions in nanopores,
    84–89
excess adsorption/desorption
    isotherms, 78–81, 80*f*
GCMC-based approach *vs.*
    conventional approach, 98–101,
    99–100*f*
grand canonical Monte Carlo (GCMC)
    simulations, 81–84
pore-size distributions, 78, 79*f*, 87–88,
    87*f*
shale sample, characterization of,
    76–78, 76*t*, 77*f*, 79*f*
system pressure, effect of, 84–85, 85*f*
system temperature, effect of, 85–87,
    86*f*
TOC and BET surface areas, 76*t*
confined C₁–*n*C₆ mixture, phase behavior
    of, 22–35, 24–27*f*, 29–32*f*, 34*f*
confined pure *n*C₈, phase behavior and
    critical properties of, 16–22, 17–20*f*,
    22*f*
density functional theory (DFT), 11–13,
    12*t*
dew-point calculation, 15
molecular model and theory, 11–15
N₂/*n*-C₄H₁₀, phase behavior of, 35–52
bubble-point pressure, 45–47, 46*t*
compositions and molar numbers,
    40–41, 41*t*
mass, pore volume, and TOC of shale
    samples, 39, 40*t*
in partially confined space, 44–47,
    44–46*f*
pore size distribution, 43–44, 43*f*
P/V isotherms, measurement of,
    38–42, 39*f*, 44–45, 44–46*f*

sorption of individual components,
    47–49
    TOC on sorption capacity, 49–52, 50*t*,
        51*f*
PR-EOS with capillary pressure model,
    13–15
pure hydrocarbons, critical properties of,
    15–16
Simplified local density (SLD) model,
    106–111
    absolute adsorption of CH₄ on, 106–111,
        118–121, 118–119*t*
    GCMC simulations, density profile
        comparison, 143
    validation, 115–118, 116*f*
"Sorbed gas" reservoirs, 47–48
Sorption capacity
    defined, 49
    TOC effect on, 49–52, 50*t*, 51*f*
Standard deviation (SD), 174–176
Successive substitution (SS), 14–15
Supercritical CO₂, 167

T

Temperature-dependent X-ray diffraction
    technology, 36–37
Thermogravimetric analyzer (TGA), 78–79,
    101–102, 122–123
Total organic carbon (TOC), N₂/*n*-C₄H₁₀
    phase behavior
    shale sample, 39, 40*t*, 44
    sorption capacity, 49–52, 50*t*, 51*f*
Transition zone, 211–213

V

van der Waals constant, 75, 81, 90–92,
    98
Vapor-liquid density, 16
Volume shift parameter (VSP), 12
Volumetric method, 101–102

W

Weighted-density approximation (WDA),
    11–12

Y

Young-Laplace (YL) equation, 2, 9–11

Printed in the United States
by Baker & Taylor Publisher Services